# SpringerBriefs in Physics

SpringerBriefs in Physics are a series of slim high-quality publications encompassing the entire spectrum of physics. Manuscripts for SpringerBriefs in Physics will be evaluated by Springer and by members of the Editorial Board. Proposals and other communication should be sent to your Publishing Editors at Springer.

Featuring compact volumes of 50 to 125 pages (approximately 20,000–45,000 words), Briefs are shorter than a conventional book but longer than a journal article. Thus, Briefs serve as timely, concise tools for students, researchers, and professionals.

Typical texts for publication might include:

- A snapshot review of the current state of a hot or emerging field
- A concise introduction to core concepts that students must understand in order to make independent contributions
- An extended research report giving more details and discussion than is possible in a conventional journal article
- A manual describing underlying principles and best practices for an experimental technique
- An essay exploring new ideas within physics, related philosophical issues, or broader topics such as science and society

Briefs allow authors to present their ideas and readers to absorb them with minimal time investment. Briefs will be published as part of Springer's eBook collection, with millions of users worldwide. In addition, they will be available, just like other books, for individual print and electronic purchase. Briefs are characterized by fast, global electronic dissemination, straightforward publishing agreements, easy-to-use manuscript preparation and formatting guidelines, and expedited production schedules. We aim for publication 8–12 weeks after acceptance.

More information about this series at http://www.springer.com/series/8902

Walter Dittrich

# The Development of the Action Principle

## A Didactic History from Euler-Lagrange to Schwinger

 Springer

Walter Dittrich
Institute of Theoretical Physics
University of Tübingen
Tübingen, Baden-Württemberg, Germany

ISSN 2191-5423           ISSN 2191-5431   (electronic)
SpringerBriefs in Physics
ISBN 978-3-030-69104-2       ISBN 978-3-030-69105-9   (eBook)
https://doi.org/10.1007/978-3-030-69105-9

This Springer imprint is published by the registered company Springer Nature Switzerland AG
The registered company address is: Gewerbestrasse 11, 6330 Cham, Switzerland

*This volume is dedicated to my dear wife Ginny Dittrich.*

# Preface

This work is devoted to the development of the history of the principle of stationary action. It will be divided into two parts: in the first, we will pay tribute to the founders of the action principle, while the second part provides many different worked-out, selected examples which cover in great detail the achievements between the seventeenth and the beginning of the twentieth century. It is exciting to see how the greatest scientists of their time struggled with a final, mathematically satisfying formulation. Foremost Daniel Bernoulli, Euler, d'Alembert and Laplace shaped the history of the early development of the action principle. Later on, Lagrange, Hamilton and last, not least, Schwinger finalized and polished the early attempts to transform almost all of nature's formidable problems into one dynamical principle.

How this principle is put to work and how much we have learned since Euler's fundamental contributions to the lemniscate problem (which spurred Gauß' motivation to develop the elliptic functions) and his introduction of the first classical field theory, namely, of the mathematically founded theory of hydrodynamics, is the leading theme in the present work.

In the nineteenth and twentieth centuries, we see the action principle celebrating triumphs in all branches of classical and quantum theory. The discovery of Einstein's field equations in gravity by Hilbert via the action principle and all the other applications of the action principle in classical and quantum theories of modern times can be traced back to the magnificent achievements by Euler and his contemporaries, and are valid still today.

Tübingen, Germany
April 2021

Walter Dittrich

# Acknowledgements

My sincerest thanks go to Ms. Ute Heuser, who has guided almost all of my book publications with great commitment through Springer's sometimes challenging publication procedures. The huge world-wide proliferation of our—my and my co-authors'—books can be credited in no small part to her. Needless to say, it is always a pleasure to speak with her on the phone or to read her "good news" announcing the acceptance of a new manuscript for publication. *Ad multos annos* is my greatest wish.

# Short Historical Introduction

The first chapter of this history is primarily characterized by the study of the *curva elastica*. It was the curva elastica that was the starting point of one of the most exciting developments in the history of mathematics. The functional representation of this curve remained obscure for a long period, since it could not be expressed with functions known at the time.

In particular, the *curva elastica* had aroused strong emotions and led to a rift between the Bernoulli brothers due to hefty priority conflicts. Johann went so far as to publically announce that he would not return to Basel until his brother, Jakob, was dead—the same older brother who had introduced him to the secrets of mathematics. Jakob died in 1705, but his family did not allow Johann access to his brother's inheritance. Before his death, Jakob had, however, made great progress regarding the *curva elastica*. He succeeded in reducing the problem in the rectification of the *curva elastica* to the so-called *lemniscate*. This Bernoullian *lemniscate* is a special case of the Cassini curves and looks like the mathematical symbol for infinity. The name comes from the Greek $\lambda \varepsilon \mu \nu i \sigma \kappa o \zeta$, meaning a ribbon or band. Count Fagnano achieved an equally important breakthrough in the same direction. By using the doubling equation for the *lemniscate* arc, he made a contribution of lasting value. Self-confidently, he had the title page of this work decorated with a *lemniscate* arc. Fagnano had sent his research that had been published in 1750 to the Berlin Academy, of which he had previously become a member. These works landed in the hands of Euler on December 23, 1750. The details were passed on by Jacobi, who, in preparing an edition of Euler's complete works, wrote: *On this occasion I also discovered an unusually important day in the history of mathematics, [the day] on which our academy requested Euler to examine the work that had been submitted by Fagnano before acceptance for publication. As a result of Euler's investigation, the theory of elliptic functions emerged.* Of all Euler's discoveries, the addition theorem for *lemniscate* integrals and their generalization had the greatest impact on the further development of analysis in the nineteenth century.

Nevertheless, it should be said that in spite of the brilliant results of Euler's work on the *lemniscate* produced, he was still dealing with elliptic or, better, *lemniscate* or Fagnano integrals, and not elliptic functions, which are the inverse of elliptic integrals. The latter were first introduced in the history of mathematics by Gauß.

Just as the history of elliptic integrals began with the lemniscate integral, so did the history of the doubly periodic *lemniscate* or of the elliptic functions start with their inversion and continuation into the complex plane. Their beginnings and the developments that followed explosively already emerged during Gauß' lifetime in the work done by Abel, Jacobi, Eisenstein, Weierstraß, etc., and were in fact due to Gauß' further strokes of genius. Unfortunately, he never published anything on this success in the realm of elliptic functions.

# Contents

# About the Author

Author (r) with J. Schwinger, 1981. Photo: Tübingen University, Germany

**Prof. Dr. Walter Dittrich** was head of the quantum electrodynamics group at the University of Tübingen until his retirement in 2001 and is still actively publishing papers and books in classical and quantum physics. He received his doctorate under Prof. Heinz Mitter at Heisenberg's institute in Munich and continued pre- and postdoc work at Brown University, Harvard and MIT. He profited immensely from lectures by and discussions with Profs. Herb Fried, Ken Johnson, Steve Weinberg, Julian Schwinger and, later on, at the Institute for Advanced Study (IAS) in Princeton, Steve Adler and David Gross at Princeton University. He started his work on gauge theories and QED in collaboration with Schwinger in the late 1960s. He was visiting professor at UCLA, Berkeley, Stanford and the IAS. He has over 30 years of teaching experience and is one of the key scientists in developing the theoretical framework of quantum electrodynamics.

# Chapter 1
# Curva Elastica

In the following, we will consider the problem of determining the forms which an infinitesimally thin rod can take when held by constant forces at the end only. A measure for the rod's bending is denoted by the flexural rigidity $B$. (If the rod had a finite cross section, we would find for $B$ the product $EI$, where $E$ is Young's modulus for the material of the wire, and $I$ is its moment of inertia.) We are dealing instead with a line - the neutral line -, which is neither stretched nor compressed; so the rod is in equilibrium under the action of the force couple $F$ and $-F$, i.e., the force acting as a couple on the curve line is zero (See Fig. 1.1.).

The bending moment (torque) at the point $P(x, y)$ of the curve is a flexural couple $D = D\hat{z}$ in the $z$ direction at $(x, y)$ with magnitude

$$D = D_z = F \cdot y, \qquad (1.1)$$

where $F$ is constant all along the arc, and is, in fact, the magnitude of the single force in the string which connects the two endpoints on the $x$ line. $y$ is the distance from the $x$ axis.

If $\rho$ is the radius of curvature at the point $(x, y)$ in a Cartesian coordinate system where the $x$ axis is horizontal, the $y$ axis is vertical, and radius $\rho$ of curvature positive when the curve is concave upward ($\cup$), it follows from calculus (Newton) that ($y' = dy/dx$):

$$\frac{1}{\rho} = \frac{y''}{(1 + y'^2)^{\frac{3}{2}}} . \qquad (1.2)$$

However, since we want the bending moment positive when the curve is concave downward ($\cap$), we have to replace $\rho$ by $-\rho$:

© The Author(s), under exclusive license to Springer Nature Switzerland AG 2021
W. Dittrich, *The Development of the Action Principle*,
SpringerBriefs in Physics,
https://doi.org/10.1007/978-3-030-69105-9_1

**Fig. 1.1** Curva elastica

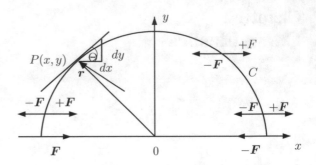

$$\frac{1}{\rho} = \frac{-y''}{(1+y'^2)^{\frac{3}{2}}} \, .$$

(1.3)

Writing

$$D = F \cdot y = B\frac{1}{\rho} = B\kappa \, , \qquad y = \frac{B}{F}\frac{1}{\rho} \, ,$$

(1.4)

we see that the curvature $\kappa$ is proportional to the coordinate $y$ at the point $(x, y)$: $\kappa \sim y$. At this stage it is convenient to introduce the substitution $a^2 = \frac{4B}{F}$ so that

$$y = \frac{a^2}{4}\frac{1}{\rho} = \frac{a^2}{4}\kappa \, .$$

(1.5)

Then Eq. (1.3) together with (1.2) turns into

$$y = \frac{-a^2}{4}\frac{y''}{(1+y'^2)^{\frac{3}{2}}} \, .$$

(1.6)

Note: It took several years from the beginning of the 18th century for giants like Leibniz, Jacob and Johann Bernoulli, as well as Euler, to discover and then understand the full meaning of Eq. (1.6). Also Daniel Bernoulli and Fagnano delivered important contributions. The highpoint in solving the equation for the *elastica* was reached when Gauß (in January of 1797, when he was just 19 years old) turned his attention to the *lemniscate* problem. This was one of the brilliant moments in the history of mathematics and theoretical physics. Gauß noticed that the *curva elastica (lemniscata)* opened a window on a brand new field of mathematics when he introduced the a.g.M. (arithmetic-geometric-mean or elliptical functions). He also extended the new functions into the complex plane! The followers, Legendre, Weierstrass, Jacobi and many others, extended and completed the whole problem which began with such a simple physical object as the *curva elastica*. We will cast a brief glance at the achievements that were made over the centuries, but which are still of great importance up to the present.

By the way, in the first articles by the Bernoullis and Euler the coordinates $(x, y)$ were interchanged to $(y, x)$ so that they arrived at a slightly modified form of (1.6); their starting point was therefore

$$x = -\frac{a^2}{4} \frac{d^2x/dy^2}{\left(1 + \left(\frac{dx}{dy}\right)^2\right)^{3/2}}. \tag{1.7}$$

In their notation we would then find

$$D = xF \qquad \rho = \frac{1}{\kappa} \qquad x = \frac{a^2}{4}\kappa, \qquad i.e. \qquad \kappa \sim x.$$

Originally we reproduced their result starting with Eq. (1.7). Of course the final results are identical, but since modern convention favors $x$ for the horizontal axis, and $y$ for the vertical axis, we now continue with Eq. (1.6).

Now let $\Theta$ be the angle which the tangent to the curve makes with the $x$ axis. Then

$$\frac{dx}{ds} = \cos\Theta, \qquad \frac{dy}{ds} = \sin\Theta, \qquad \frac{dy}{dx} = y' = \tan\Theta,$$

$$1 + y'^2 = 1 + \tan^2\Theta = \frac{1}{\cos^2\Theta}, \qquad \sqrt{1 + y'^2} = \frac{1}{\cos\Theta} = \sec\Theta,$$

$$y(x) = -\frac{a^2}{4} \frac{y''}{(1 + y'^2)^{3/2}}.$$

On multiplying both members of this relation by $2y'$ and then integrating, there results

$$y^2 = C + \frac{a^2}{2} \frac{1}{(1 + y'^2)^{1/2}}, \tag{1.8}$$

where $C$ is the constant of integration. With the aid of (1.6) the above Eq. (1.8) can be written as

$$y^2 = C + \frac{1}{2}a^2 - a^2 \sin^2\frac{\Theta}{2} = h^2 - a^2 \sin^2\frac{\Theta}{2}, \tag{1.9}$$

where

$$h^2 = C + \frac{1}{2}a^2.$$

Further integration has to distinguish between the relative values of $h$ and $a$, namely

$$h^2 < a^2 \qquad h^2 = a^2, \qquad h^2 > a^2.$$

(a) The equation of the elastic curve for $h^2 < a^2$. Let

$$h^2 = k^2 a^2 (k^2 < 1) ; \qquad -\sin \frac{\Theta}{2} = k \sin \varphi .$$

Then

$$y = h \cos \varphi . \tag{1.10}$$

We also need

$$\sin \Theta = \frac{dy}{ds} = \frac{dy}{d\varphi} \frac{d\varphi}{ds} = -h \sin \varphi \frac{d\varphi}{ds} ,$$

so that

$$\frac{ds}{d\varphi} = -h \frac{\sin \varphi}{\sin \Theta} . \tag{1.11}$$

From the relation $-\sin \frac{\Theta}{2} = k \sin \varphi$ we find

$$-\sin \Theta = -\sqrt{1 - \cos^2 \Theta} = -2 \sqrt{\frac{1 - \cos \Theta}{2}} \sqrt{\frac{1 + \cos \Theta}{2}} = -2 \sin \frac{\Theta}{2} \cos \frac{\Theta}{2}$$

$$= -2 \sin \frac{\Theta}{2} \sqrt{1 - \sin^2 \frac{\Theta}{2}} = -2 \sin \frac{\Theta}{2} \sqrt{1 - k^2 \sin^2 \varphi}$$

$$-\sin \Theta = 2k \sin \varphi \sqrt{1 - k^2 \sin^2 \varphi} .$$

Hence, we obtain from (1.11)

$$\frac{ds}{d\varphi} = \frac{a}{2} \frac{1}{\sqrt{1 - k^2 \sin^2 \varphi}}$$

and when integrated

$$s = \frac{a}{2} \int_0^\varphi \frac{d\varphi}{\sqrt{1 - k^2 \sin^2 \varphi}} = \frac{a}{2} F(k, \varphi) ; \tag{1.12}$$

here $F(k, \varphi)$ is Legendre's elliptic integral of the first kind, which is derived from its normal form

$$F(k, x) = \int_0^x \frac{dt}{\sqrt{(1 - t^2)(1 - k^2 t^2)}} , \qquad k^2 < 1 ,$$

when setting $t = \sin \varphi$, i.e., $d\varphi = \frac{dt}{\sqrt{1-t^2}}$. From

$$\cos \Theta = \frac{dx}{ds}$$

follows

$$\frac{dx}{ds} = \frac{ds}{d\varphi} \cos \Theta = -h \frac{\sin \varphi}{\tan \Theta}$$

$$= \frac{1}{2} a \frac{1 - 2k^2 \sin^2 \varphi}{\sqrt{1 - k^2 \sin^2 \varphi}} \, ;$$

therefore

$$x = a \int_0^\varphi \sqrt{1 - k^2 \sin^2 \varphi} \, d\varphi - \frac{a}{2} \int_0^\varphi \frac{d\varphi}{\sqrt{1 - k^2 \sin^2 \varphi}}$$

$$= a E(k, \varphi) - \frac{a}{2} F(k, \varphi) , \qquad (1.13)$$

where $E(k, \varphi)$ is Legendre's elliptic integral of the second kind with normal form

$$E(k, x) = \int_0^x \sqrt{\frac{1 - k^2 t^2}{1 - t^2}} \, dt .$$

In combination of Eq. (1.13) for $x$ with the equation

$$y = h \cos \varphi$$

these parametric equations provide us with the following Figs. 1.2, 1.3, 1.4, 1.5 and 1.6.

(b) Equation of the elliptic curve for $h^2 = a^2$.

In this case since the constants $h$ and $a$ are equal to the parameter $k = 1$, the solution curve does not lead to elliptic integrals. In fact, the solution in that case is expressible in terms of elementary functions. Thus,

**Fig. 1.2** Curva elastica with $k = \sin 30°$

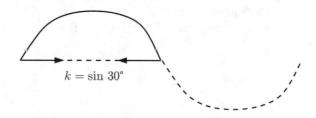

$$k = \sin 30°$$

**Fig. 1.3** Curva elastica with
$k = \sin 45°$

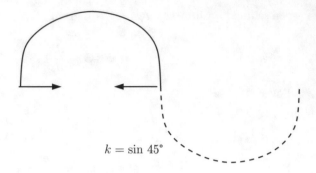

$k = \sin 45°$

**Fig. 1.4** Curva elastica with
$k = \sin 55°$

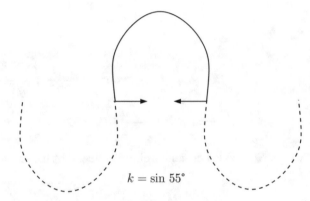

$k = \sin 55°$

$$y = a \cos \varphi,$$

$$x = a \int_0^{\varphi} \left( \cos \varphi - \frac{1}{2} \frac{1}{\cos \varphi} \right) d\varphi = a \sin \varphi - \frac{a}{2} \log \tan \left( \frac{\varphi}{2} + \frac{\pi}{4} \right),$$

and

$$s = \frac{a}{2} \int_0^{\varphi} \frac{d\varphi}{\cos \varphi} = \frac{a}{2} \log \tan \left( \frac{\varphi}{2} + \frac{\pi}{4} \right).$$

Since $x$ tends forward infinity as $\varphi$ approaches $\frac{\pi}{2}$, the complete curve has but a single loop the two branches of which are asymptotic to the $x$ axis (Fig. 1.7).

(c) The equation of the elastic curve for $h^2 > a^2$.

When the substitution $a^2 = k^2 h^2 (k^2 < 1)$ is made, it is found that

$$y = h\sqrt{1 - k^2 \sin^2 \frac{\theta}{2}},$$

**Fig. 1.5** Curva elastica with
$k = \sin 65° 22'$

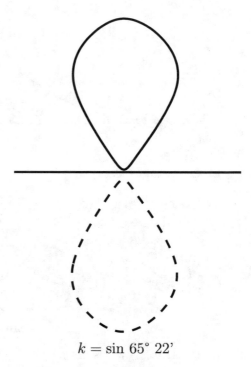

$$k = \sin 65° \; 22'$$

**Fig. 1.6** Curva elastica with
$k = \sin 75°$

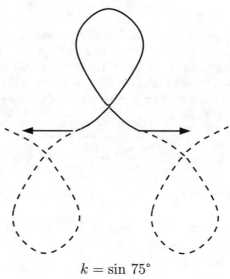

$$k = \sin 75°$$

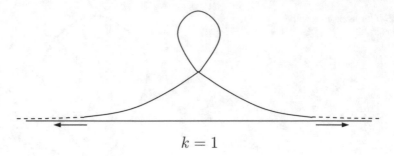

$$k = 1$$

**Fig. 1.7**  Curva elastica with $k = 1$

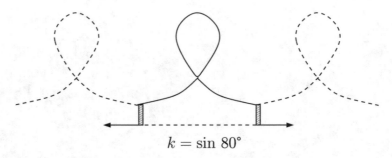

$$k = \sin 80°$$

**Fig. 1.8**  Curva elastica with $k = \sin 80°$

from which it is evident that $y$ never vanishes. Now remember that $\kappa \sim y$, so that the curvature vanishes only with $y$ and it becomes clear that there are no points of reflection on the curve. The radius of curvature always is finite and of the same sign and the tangent of the curve turns always in the same direction (Fig. 1.8).

On setting $\sin \varphi = -k \sin \frac{\theta}{2}$ and carrying through an analysis similar to that in (a) it is found that

$$y = h\sqrt{1 - k^2 \sin^2 \frac{\theta}{2}}, \qquad s = \frac{1}{2}hk^2 \, F\left(k, \frac{\theta}{2}\right)$$

and

$$x = hE\left(k, \frac{\theta}{2}\right) - h\left(1 - \frac{1}{2}k^2\right) F\left(k, \frac{\theta}{2}\right).$$

# Chapter 2
# The Curva Elastica, a Curve of Least Energy

Before we start with the variational approach of the *curva elastica* to find its equilibrium state, i.e., with minimal curvature, we want to introduce still another way to parametrize the *elastica* arc.

So far, we have dealt with Cartesian coordinates that helped us in setting up the differential equation which is to be satisfied by the curve:

$$y'' = -\frac{4}{a^2}(1 + y'^2)^{3/2}y.$$

In the foregoing section we gave various analytical and graphical solutions for the various elastic curves. In particular, we formulated the curvature in the form

$$\kappa = -\frac{y''}{(1 + y'^2)^{3/2}} \left(= -\frac{1}{\rho}\right)$$

and the associated arc length by

$$ds = \sqrt{1 + y'^2}dx.$$

We are now interested in minimizing the energy functional $E\{\kappa(\cdot)\}$, which in Cartesian coordinates is given by

$$E\{\kappa(\cdot)\} = \int \frac{(y'')^2}{(1 + y'^2)^{5/2}}dx. \tag{2.1}$$

However, instead of the independent variable $x$ we can also use the arc length as independent variable. Then the energy functional can be expressed in an equivalent form.

© The Author(s), under exclusive license to Springer Nature Switzerland AG 2021
W. Dittrich, *The Development of the Action Principle*,
SpringerBriefs in Physics,
https://doi.org/10.1007/978-3-030-69105-9_2

To derive this expression, let us start with $t$ as parameter for the evolving curve from $t_0$ to $t$:

$$s(t) = \int_{t_0}^{t} \sqrt{\left(\frac{dx}{dt}\right)^2 + \left(\frac{dy}{dt}\right)^2}\, dt \,.$$

Inverting $s(t)$ gives $t(s)$, with

$$\frac{dt}{ds} = \acute{t}(s) = \frac{1}{\dot{s}(t)} = \frac{1}{\sqrt{\dot{x}^2 + \dot{y}^2}} \,.$$

Here we have introduced $\dot{x} = \frac{dx}{dt}$, where $\dot{} = \frac{d}{dt}$ and $' = \frac{d}{ds}$.

Now we are going to use $s$ instead of the $t$ parameter:

$$x(t) = x(t(s)) =: \bar{x}(s) \,,$$
$$y(t) = y(t(s)) =: \bar{y}(s) \,.$$

It then follows that

$$\bar{x}'(s) = \dot{x}(t)\acute{t}(s) = \frac{\dot{x}}{\sqrt{\dot{x}^2 + \dot{y}^2}} = \cos\vartheta \,,$$

$$\bar{y}'(s) = \frac{\dot{y}}{\sqrt{\dot{x}^2 + \dot{y}^2}} = \sin\vartheta \,,$$

and therefore

$$\bar{x}'^2(s) + \bar{y}'^2(s) = 1 \,;$$

in short, by replacing incorrectly $\dot{}$ by $'$

$$\dot{x}^2(s) + \dot{y}^2(s) = 1 \,.$$

From our results

$$\dot{x}(s) = \cos\vartheta(s) \,,$$
$$\dot{y}(s) = \sin\vartheta(s) \,,$$
$$\dot{x}^2(s) + \dot{y}^2(s) = 1 \,,$$

we find

$$\ddot{x}^2 + \ddot{y}^2 = \left(\frac{d}{ds}\dot{x}\right)^2 + \left(\frac{d}{ds}\dot{y}\right)^2 = \left(\frac{d}{ds}\cos\vartheta\right)^2 + \left(\frac{d}{ds}\sin\vartheta\right)^2$$
$$= (-\sin\vartheta\dot{\vartheta})^2 + (\cos\vartheta\dot{\vartheta})^2 = (\sin^2\vartheta + \cos^2\vartheta)\dot{\vartheta}^2$$
$$= \dot{\vartheta}^2 .$$

With the definition of

$$\kappa = \dot{\vartheta} = \frac{d\vartheta}{ds}, \qquad \kappa^2 = \dot{\vartheta}^2$$

we have as an equivalent for the curvature

$$\kappa^2(s) = \ddot{x}^2(s) + \ddot{y}^2(s) . \tag{2.2}$$

Hence, besides (2.1) we have another choice to minimize the energy functional

$$E\{\kappa(\cdot)\} = \int \kappa^2(s)ds = \int (\ddot{x}^2 + \ddot{y}^2)ds \tag{2.3}$$

with the constraint

$$\dot{x}^2 + \dot{y}^2 = 1 . \tag{2.4}$$

When we vary (2.1) we should work with the constraint that the total arc length is given some prescribed value. So we need a Lagrangian multiplier $\lambda$. The introduction of this constraint does not complicate the calculation very much. Nevertheless, we will assume that such a constraint is not prescribed and the constant Lagrangian multiplier is put equal to zero. Hence, it will suffice to neglect the $\lambda$-term in the variation

$$\delta \int \left\{ y''^2(1 + y'^2)^{-5/2} + \lambda(1 + y'^2)^{1/2} \right\} dx = 0 . \tag{2.5}$$

The other alternative requires us to minimize (2.3), i.e., to perform the variation

$$\delta \int \left\{ (\ddot{x}^2 + \ddot{y}^2) + \mu(s)(\dot{y}^2 + \dot{x}^2) \right\} ds = 0 ,$$

where the $s$-dependent Lagrangian multiplier $\mu(s)$ is introduced and will be determined to obtain $\dot{x}^2 + \dot{y}^2 = 1$.

I have worked out both variational problems, but shall here—for the sake of simplicity—treat only (2.5) in detail, with $\lambda = 0$. At the end of this section, I shall, however, present the full solution.

Between 1680 and 1690, Newton's and Leibniz's differential and integral calculus became thoroughly understood. Thereafter, in 1744, Euler, as a member of the St. Petersburg Academy of Science, published his "*Methodus inveniendi curvas lineas*"

[1]. Euler continued the pioneering work done a century earlier by the Bernoulli brothers Jakob and Johann, but it was doubtless Jakob's nephew Daniel, who ignited Euler's analysis of the *elastica*. The breakthrough occurred in 1742, when Daniel proposed to Euler solving the general *elastica* problem with the new technique of generating a whole set of curves from an elastic rod with given length and arbitrary given tangent constraints at the endpoints [2].

Instead of the rod, we now consider an elastic, flexible wire which, under the influence of its own internal tension, assumes a certain equilibrium position. The functional form of this tension balance is to be determined. For this, we assume that the internal tensions possess a certain potential energy. Equilibrium occurs when the potential energy of the system is stationary under the action of constraints, whereby the equilibrium is stable when the energy functional assumes a minimum.

Euler (who else?) was the first to build upon (and mention) the work of the Bernoullis, and to fully characterize the curve family known as the *elastica* [3]. This work was published in the appendix of his trailblazing book on variation techniques. His treatment was very effective and still meets modern standards.

Following an idea of Daniel Bernoulli, Euler expressed the *elastica* problem very clearly in variational form. He wrote (translated from Latin into English):

"Among all curves of equal length which not only pass through the points $A$ and $B$, but are also tangent to certain lines at these points, it is the one which minimizes the value of the expression $\int \frac{ds}{RR}$."

Here, $s$ refers to the *arc* length, just as is customary today, and $R$ is the radius of curvature (in Euler's words, "*radius osculi curvae*"), equal to $\frac{1}{\kappa}$ in modern notation. Therefore, we now define in more detail that the *elastica* is the curve that minimizes the energy functional $E\{\kappa(\cdot)\}$ over the length of the curve.

Euler employed Cartesian coordinates, using the standard definitions of $ds = \sqrt{1 + y'^2}dx$ and $\frac{1}{R} = \frac{y''}{(1+y'^2)^{\frac{3}{2}}}$, where $y'$ and $y''$ represent $\frac{dy}{dx}$ and $\frac{d^2y}{dx^2}$, respectively. Thus, the variation problem is employed to determine the minimum for

$$\int \frac{y''^2 dx}{(1 + y'^2)^{\frac{5}{2}}} .$$

This equation contains the first and second derivatives of $y(x)$, so that the simple Euler–Lagrange equation is insufficient. Daniel Bernoulli encountered this difficulty, as he expressed in his letter of 1742. In hindsight, we know that the formulation of the problem with respect to the tangent angle as a function of the arc length leads to a variation of the first derivative, but apparently this was unclear to both Bernoulli and Euler at the time.

We ought to mention that around the same time there was another problem of interest in the air. It was the *brachistochrone* problem that was posed by Johann Bernoulli in 1696. He already knew that the solution was a cycloid (discovered by Huygens in 1659). The *brachistochrone* problem solved by Jakob Bernoulli (1697) was the most profound, because it recognized the "variable curve" aspect of the problem, and it is now considered by some as the first major step in the development of the calculus of variation. There is, however, a huge difference in minimizing a certain definite integral and solving the following Euler–Lagrange equation. It is for the *brachistochrone* $y'\frac{\partial f}{\partial y'} - f = const.$, $y' = \frac{dy}{dx}$ and where $f(y, y')$ does not contain $x$ explicitly. The function $f$ is given by $\frac{\sqrt{1+y'^2}}{\sqrt{k-y}}$, $k = const..$

As is well known, the solution of this equation can be parameterized by equations containing the *sin* and *cos* functions. And here is the main difference: no new type of functions were necessary besides the familiar ones. The *elastica (lemniscate)*, on the other hand, opened the door to a completely novel type of functions and integrals: the elliptic functions and integrals. The function to be minimized for the *lemniscate* is now given by $f(y', y'') = \frac{y''^2}{(1+y'^2)^{\frac{5}{2}}}$.

Also the Euler–Lagrangian looks totally different: $\frac{\partial f}{\partial y} - \frac{d}{dx}\frac{\partial f}{\partial y'} + \frac{d^2}{dx^2}\frac{\partial f}{\partial y''} = 0$, which Daniel Bernoulli couldn't solve but Euler did.

Now, we want to retroactively determine the curves for which

$$E\{\kappa(\cdot)\} = \int \kappa^2(x)dx \qquad (2.6)$$

becomes minimal. With the well-known formulas

$$\frac{1}{\rho} = \kappa = \frac{y''}{(1+y'^2)^{\frac{3}{2}}} \quad \text{and} \quad ds = \sqrt{1+(y')^2}dx$$

we obtain

$$E\{\kappa(\cdot)\} = \int \frac{(y'')^2 dx}{[1+(y')^2]^{\frac{5}{2}}} . \qquad (2.7)$$

This expression is of the form

$$\int F(x, y; y', y'')dx$$

and the calculus of variation teaches us that the stationary value of the integral is the solution of the following (Euler) differential equation

$$F_y - \frac{d}{dx}F_{y'} + \frac{d^2}{dx^2}F_{y''} = 0, \qquad (2.8)$$

where $F_y$, $F_{y'}$, $F_{y''}$ are the partial derivatives with regard to $y$, $y'$ and $y''$. The individual derivatives are then given by

$$F_y = 0 \,,$$

$$F_{y'} = -\frac{5y'(y'')^2}{[1+(y')^2]^{\frac{7}{2}}} \,,$$

$$F_{y''} = \frac{2y''}{[1+(y')^2]^{\frac{5}{2}}} \,.$$

Since $F_y = 0$, it follows that

$$-\frac{d}{dx}\left[F_{y'} - \frac{d}{dx}F_{y''}\right] = 0 \,,$$

so that a first integration gives

$$-F_{y'} + \frac{d}{dx}F_{y''} = A \,, \tag{2.9}$$

where $A$ is an arbitrary constant. With

$$\frac{d}{dx}F_{y''} = \frac{2y''}{[1+(y')^2]^{\frac{5}{2}}} - \frac{10y'(y'')^2}{[1+(y')^2]^{\frac{7}{2}}} \,,$$

follows

$$\frac{2y''}{[1+(y')^2]^{\frac{5}{2}}} - \frac{5y'(y'')^2}{[1+(y')^2]^{\frac{7}{2}}} = A \,.$$

Since

$$\frac{dy''}{dy'} = \frac{dy''}{dx}/\frac{dy'}{dx} = \frac{y'''}{y''} \,, \text{ i.e.,} \qquad y''' = y''\frac{dy''}{dy'} \,,$$

and with

$$\frac{d}{dy'}\frac{1}{2}(y'')^2 = y''\frac{dy''}{dy'} \,,$$

we get

$$\frac{d}{dy'}\frac{1}{[1+(y')^2]^{\frac{5}{2}}} = -\frac{5y'}{[1+(y')^2]^{\frac{7}{2}}} \,.$$

Inserted in (2.9) this result gives

$$\frac{d}{dy'}\frac{(y'')^2}{[1+(y')^2]^{\frac{5}{2}}} = A$$

and a further integration leads to

$$\frac{(y'')^2}{[1 + (y')^2]^{\frac{5}{2}}} = Ay' + B ,\tag{2.10}$$

where $B$ is a second arbitrary constant.

Hence, for a curve beginning at $(x_0, y_0)$ and ending at $(x_1, y_1)$, the integral of the square of the curvature reads

$$\int \kappa^2 ds = \int [Ay' + B]dx = \int A dy + \int B dx$$
$$= A(y_1 - y_0) + B(x_1 - x_0) .$$

Since

$$\frac{(y'')^2}{[1 + y'^2]^{5/2}} = \kappa^2 \sqrt{1 + y'^2}$$

it follows, because of (2.10)

$$\kappa^2 \sqrt{1 + y'^2} = Ay' + B ,\tag{2.11}$$

and with

$$\frac{dy}{dx} = \frac{dy}{ds} : \frac{dx}{ds} = \frac{\dot{y}}{\dot{x}} ,$$

$$\kappa^2 \sqrt{1 + \frac{\dot{y}^2}{\dot{x}^2}} = A\frac{\dot{y}}{\dot{x}} + B : \qquad \kappa^2 \sqrt{\dot{x}^2 + \dot{y}^2} = A\dot{y} + B\dot{x} , \qquad \kappa^2 = \kappa_{min}^2 .$$

Here, then, is the minimal $\kappa$ fur the *curva elastica*:

$$\kappa_{min}^2 = \frac{A\dot{y} + B\dot{x}}{\sqrt{\dot{x}^2 + \dot{y}^2}} .\tag{2.12}$$

Had we done the whole calculation with the Lagrangian multiplicator $\lambda$, we would have obtained a slight modification of (2.12), namely

$$\kappa_{min}^2 = A\frac{y'}{\sqrt{1 + y'^2}} + B\frac{1}{\sqrt{1 + y'^2}} + \lambda \tag{2.13}$$

or, with

$$y' = \frac{\dot{y}(s)}{\dot{x}(s)} , \qquad \dot{x} = \cos\varphi(s) , \qquad \dot{y} = \sin\varphi(s) , \qquad \dot{x}^2 + \dot{y}^2 = 1 ,$$

$$\kappa_{min}^2 = \frac{A\dot{y}}{\sqrt{\dot{x}^2 + \dot{y}^2}} + \frac{B\dot{x}}{\sqrt{\dot{x}^2 + \dot{y}^2}} + \lambda .\tag{2.14}$$

The $s$ parametrization would have given

$$\kappa^2_{min} = 2c_1 \dot{x} + 2c_2 \dot{y} + d \tag{2.15}$$

or

$$\kappa^2_{min} = \ddot{x}^2 + \ddot{y}^2 = \frac{2d}{3} - 4\wp\,(s + a;\, g_2,\, g_3)\,,$$

with various integration constants and $g_2, g_3$, which occur in the differential equation for the Weierstrass elliptic function $\wp$. So the squared curvature of a curve for which the integral of the squared curvature along the arc length is a minimum can be expressed by means of a Weierstrass elliptic function; the periods of the elliptic function depend on the boundary condition. This situation reminds one very much of the Legendre elliptic integrals that we met in the previous section.

The two Eqs. (2.14) and (2.15) are obviously identical in meaning, although the constants involved have different names. This becomes evident when we write for (2.14)

$$\kappa^2_{min} = A \sin\varphi + B \cos\varphi + \lambda\,,$$

which in terms of the lope angle $\varphi$ can also be written as

$$\kappa^2_{min} = \left(\frac{d\varphi}{ds}\right)^2 = r \sin(\varphi - \psi) + \gamma\,, \tag{2.16}$$

where we have introduced new constants $r$, $\psi$ and $\gamma$.

In a similar manner, we can write for (2.15)

$$\kappa^2_{min} = \left(\frac{d\varphi}{ds}\right)^2 = 2c_1 \dot{x} + 2c_2 \dot{y} + d$$
$$= 2c_1 \cos\varphi + 2c_2 \sin\varphi + d\,.$$

By rotating the original $(x, y)$ coordinate system into a new $(u, v)$ system using the (constant) rotation angle $\psi$, we find ($\kappa$ is invariant) for $\Theta = \varphi - \psi$:

$$\frac{d\Theta}{ds} = \frac{d}{ds}(\varphi - \psi) = \frac{d\varphi}{ds} = \kappa\,.$$

Now $\frac{d\Theta}{ds}$ is a measure for the "velocity" with which the tangential direction of the curve $C$ changes, i.e., with which the tangent rotates. So with respect to the $(u, v)$ coordinate system, (2.16) becomes

$$\kappa^2_{min} = \left(\frac{d\Theta}{ds}\right)^2 = r \sin\Theta + \gamma\,. \tag{2.17}$$

Noticing that $ds = \frac{du}{\cos\Theta}$ and $\sin\Theta = \cos\Theta \frac{dv}{du}$, Eq. (2.17) can be written as

$$\frac{\cos \Theta d\Theta}{\sqrt{r \sin \Theta + \gamma}} = \pm du\,.$$

After integration, this equation gives

$$\kappa_{min} = \sqrt{r \sin \Theta + \gamma} = \pm \frac{r}{2} u + \delta\,, \tag{2.18}$$

where $\delta$ is an integration constant.

The angle $\Theta$ denotes the angle between the $u$ axis and the tangent of the curve (at the point $(u, v)$). Also notice that the curvature is a linear function of $u$. This result is in agreement with our former result in the first section.

Hence we can formulate our findings as a theorem:

For a curve for which the integral along the arc length of the squared curvature is minimized, there is always a direction along which the curvature varies linearly.

The solution of (2.18) follows the standard procedure, not unlike the one of the former section, and ends with the elliptic integrals—more precisely, with a linear combination of the Legendre functions $E$ and $F$. These functions were also involved in our special example of the rod whose ends are tied together by constant forces $\pm F$; in Fig. 1.3, one can find the corresponding picture of that elastic curve for $k = 45°$, which served as our introductory example.

Altogether, with the final result (2.18), we have succeeded in a complete integration of the Euler equation for the most general *curva elastica*. In possession of this solution, we are now able to solve Leibniz's problem for the *isochrona* and *isochrona paracentrica*, i.e., the finding of a curve along which a point particle in presence of gravity descends in equal times through equal distances [4]. Furthermore, the Italian Giulio Carlo Fagnano's impressive calculations of the *curva lemniscata* were rediscovered and the circumference of the ellipse determined. This is the origin of the name elliptic integral (function). It is known that Jacobi marked the 23rd of December, 1751 as the birthday of the elliptic functions, the day on which Euler promised the Berlin Academy to report on the immortal works of Fagnano.

Then came January of the year 1797, when the 19-year-old Gauss took on the *elastica* problem. He immediately crossed out the word *elastica* and replaced it by the word *lemniscata*. His research in this field opened the door to a completely new realm of mathematics [5]. Especially his extension of the elliptic function into the complex plane was a revolutionary step. The subsequent works of Eisenstein, Weierstraß, Jacobi and others culminated in a period that brought mathematics of the 19th century to an all-time high.

Needless to say, all these discoveries became indispensable tools for solving myriad problems in physics and the engineering world. Euler's early theory of a bending beam is still one of the basics in the construction of buildings and bridges, e.g., the Eiffel Tower; and these many applications emerged from the investigation of a highly simple object called *curva elastica* [6]. Let us not forget: all this occurred immediately after the 17th century, when the calculus of Newton and Leibniz had just been invented and tested, an indispensable discovery in the history of mankind.(It is remarkable that Max Born, in his PhD thesis, investigated the elastica problem much later, in 1906. See [7].)

# References

1. Euler, L.: Chapter Additatmentum 1. Methodus inveniendi lineas curvas maximi minimive proprietate gaudentes, sive solutio problematis isoperimetrici lattissimo sensu accepti, E065 (1744). www.Eulerarchive.org
2. Bernoulli, D.: The 26th letter to Euler. Correspondence Mathématique et Physique, vol. 2. P.H. Fuss (1742)
3. Bernoulli, J.: Quadratura curvae, e cujus evolutione describitur inflexae laminae curvature. Die Werke von Jakob Bernoulli, pp. 223–227. Birkhäuser (1692). Med. CLXX; Ref. UB: L Ia 3 211-212
4. Leibniz, G.W.: De linea isochrona, in qua grave sine acceleratione descendit,.... Acta eruditorum Aprilis (1689), pp. 195–198, Mathem. Schriften Bd. V, pp. 234–237
5. Dittrich, W: C.F. Gauß und die Anfänge der elliptischen Integrale und elliptischen Funktionen. Gauß-Gesellschaft E.V: Göttingen Mitteilungen Nr. 45, pp. 93–115 (2008)
6. Bernoulli, J., Euler, L.: Abhandlungen über das Gleichgewicht und Schwingungen der ebenen elastischen Kurven. Oswalds Klassiker der Exakten Wissenschaften Nr. 175, Verlag von Wilhelm Engelmann, Leipzig (1910)
7. Born, M.: Untersuchungen über die Stabilität der elastischen Linie in Ebene und Raum, unter verschiedenen Grenzbedingungen. PhD thesis, University of Göttingen (1906)

# Chapter 3
# From Euler to Lagrange

The calculus of variation developed its full power with the appearance of Joseph-Louis Lagrange (1736–1813). In 1755 Lagrange began working on some problems which Euler and others were trying to solve at the time. The following year he applied the calculus of variation to mechanics and showed—by employing the principle of least action—that this offered a general procedure for solving dynamical problems. He forwarded his results to Euler, who was greatly impressed and withheld a paper of his own from publication, a gesture that was typical of Euler throughout his life.

This took place at the time Euler was director of the mathematics section of the Prussian Academy in Berlin. Initially, it was the great French scientist Pierre Maupertuis who presided over the academy. Following his death in 1759, Euler took over—under the supervision of the king, Frederick the Great. Since his relationship with the king became increasingly worse, Euler left Berlin in 1766 at the age of 59, much to the king's displeasure, and returned to St. Petersburg, where he died on September 18, 1783.

After Euler's departure from Berlin, Lagrange received an attractive offer from Frederick the Great, expressing the wish by the "greatest king of Europe" to have the "greatest mathematician of Europe" resident at his court.

Maybe it does not come as a surprise that Euler offered Lagrange the opportunity to join him in St. Petersburg at the still-existent Russian Academy on the Newa River. But Lagrange declined, and in 1766 became director of mathematical physics at the Berlin Academy instead.

With respect to mathematical research, the Berlin years were fruitful for Lagrange. During his twenty years at the Berlin Academy, he worked on the "Mécanique Analytique." This work was his masterpiece, a scientific poem, according to Hamilton. Unlike Euler, Lagrange remained in favor with the king.

© The Author(s), under exclusive license to Springer Nature Switzerland AG 2021
W. Dittrich, *The Development of the Action Principle*,
SpringerBriefs in Physics,
https://doi.org/10.1007/978-3-030-69105-9_3

Following the death of Frederick the Great in 1786, Paris persuaded Lagrange to become a "pensionnaire vétéran" of the Paris Academy. In 1795, he was appointed professor of mathematics at the new École Normale and later became professor at the École Polytechnique.

Although Lagrange always acted in a reserved manner toward Euler (whom he never met), it was Euler, among the older mathematicians, who influenced him most. Euler was a simple man, not given to envy. He had only a few immediate disciples, yet, as Laplace said, he was the teacher of all the mathematicians of his time.

The eighteenth century can fairly be labeled the "Age of Euler", but his influence on the development of the mathematical sciences was not restricted to that period. Still today in the 21st century, the work of many outstanding mathematicians has arisen directly from Euler. In combination with the name Lagrange, the Euler–Lagrange variational methods can be found in almost every branch of classical and quantum dynamical physics. Most of our modern elementary particle theories are based on the Euler–Lagrange action principle. They all start with the definition of an action functional which is defined by an integral over the Lagrange function.

# Chapter 4
# Laplace and the Capillary—1807

It is remarkable that the *elastica* appears as yet another solution of a fundamental physics problem—the capillary. It was Laplace who investigated the equation of the shape of the capillary in 1807.

In particular, Laplace considers the surface of a fluid trapped between two vertical plates and obtains an equation not very much different from Euler's equation (1.6), but now the sign has to be changed, since we are dealing with a curve that is concave upward:

$$y'' = \frac{4}{a^2} y (1 + y'^2)^{\frac{3}{2}} . \tag{4.1}$$

Laplace's equation, which is seen to be equivalent to (4.1), reads

$$y'' = 2(\beta y + b^{-1})(1 + y'^2)^{\frac{3}{2}} . \tag{4.2}$$

So, Laplace indeed recognized his equation as an equivalent to the *curva elastica*.

Maxwell's figure for the capillary curve is given in Fig. 1.8 of Chap. 1. It clearly shows a non-inflectional, periodic instance of the *elastica*.

To determine the capillary curve, we begin with the by now familiar *elastica* equation (with slightly changed constants):

$$\rho y = \frac{c^2}{4} , \qquad \kappa = \frac{1}{\rho} = \frac{4}{c^2} y : \qquad \kappa \sim y!, \qquad \rho = \text{radius of curvature} .$$

$$\frac{y''}{(1 + y'^2)^{\frac{3}{2}}} = \frac{4y}{c^2} , \tag{4.3}$$

W. Dittrich, *The Development of the Action Principle*,
SpringerBriefs in Physics,
https://doi.org/10.1007/978-3-030-69105-9_4

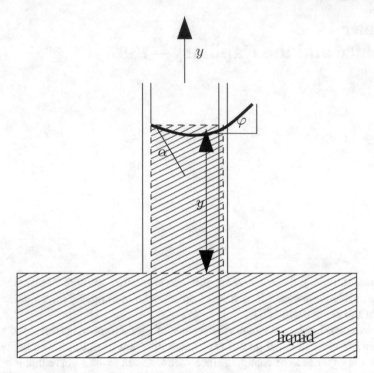

**Fig. 4.1**  The *elastica* as the surface of Laplace's capillary

$$c^2 = \frac{4\gamma}{g\sigma}, \qquad \gamma = \text{surface tension}; \; \sigma = \text{density of liquid}; \qquad g = \text{gravitational constant}.$$

Our fundamental Eq. (4.3) can easily be integrated by writing the left-hand side as

$$-\frac{1}{y'}\frac{d}{dx}\frac{1}{(1+y'^2)^{\frac{1}{2}}},$$

so that we obtain

$$-\frac{d}{dx}\frac{1}{(1+y'^2)^{\frac{1}{2}}} = \frac{4}{c^2}yy' = \frac{2}{c^2}\frac{d}{dx}(y^2),$$

which, when integrated, yields

$$\frac{2y^2}{c^2} = A - \frac{1}{(1+y'^2)^{\frac{1}{2}}} = A - \cos\varphi, \tag{4.4}$$

having used $\cos\varphi = \frac{1}{(1+y'^2)^{\frac{1}{2}}}$ from $y' = \tan\varphi$, $1 + y'^2 = 1 + \tan^2\varphi = \frac{1}{\cos^2\varphi}$.
From (4.4) follows $A - \cos\varphi > 0$ and hence $A > 1$.

Furthermore we have $\left(\rho = \frac{ds}{d\varphi}\right)$:

$$y\frac{ds}{d\varphi} = \frac{c^2}{4}.$$

Hence,

$$\frac{2\sqrt{2}}{c}\frac{ds}{d\varphi} = \frac{1}{\sqrt{A - \cos\varphi}}.$$

To bring this formula in a useful form, we put $\cos\varphi = z$ and $\frac{\sqrt{2}s}{c} = u$, and deduce

$$\frac{1}{4}\left(\frac{dz}{du}\right)^2 = (1 - z^2)(A - z). \tag{4.5}$$

The following substitution

$$z = t + \alpha,$$

where $a$ is a constant to be fixed later, brings Eq. (4.5) into the form

$$\frac{1}{4}\left(\frac{dt}{du}\right)^2 = t^3 + t^2(3\alpha - A) + t(3\alpha^2 - 2\alpha A - 1) + (\alpha^3 - A\alpha^2 - \alpha + A).$$

Here we choose $\alpha$ so that the term that multiplies $t^2$ vanishes.
Then $\alpha = \frac{A}{3}$ simplifies Eq. (4.5) enormously.
Choosing the invariants of Weierstraß's $\wp\,(u|g_2, g_3)$ function $g_2$ and $g_3$ according to

$$g_2 = \frac{4}{3}(A^2 + 3), \qquad g_3 = \frac{8A}{3}\left(\frac{A^2}{9} - 1\right) \tag{4.6}$$

we obtain the algebraic differential equation of first order,

$$\left(\frac{dt}{du}\right)^2 = 4t^3 - g_2 t - g_3. \qquad \text{(No } u \text{ on the right-hand side!)}. \tag{4.7}$$

Usually one finds in the literature the function $w = \wp(z)$, which satisfies

$$(w')^2 = \left(\frac{dw}{dz}\right)^2 = 4w^3 - g_2 w - g_3,$$

where $w'$ is an algebraic function $w$.

Finally, we have found the solution of Eq. (4.7) in form of Weierstraß's elliptic $\wp$ function,

$$t = \wp(u + \beta),\qquad(4.8)$$

where $\beta$ is another constant.

Now we choose the so far free constant $\alpha$ as the angle of capillarity (cf. Fig. 4.1). Then it is obvious that $\alpha < z = \cos\varphi < 1$ or

$$\sin\alpha - \frac{A}{3} < t < 1 - \frac{A}{3}.\qquad(4.9)$$

One of our requirements for the $\wp$ function is

$$e_1 + e_2 + e_3 = 0,$$

which can be fulfilled with $e_3 = 1 - \frac{A}{3}$, $e_2 = -1 - \frac{A}{3}$; hence, $e_1 = \frac{2A}{3}$ and $e_2 < t < e_3$.

From the definition of the doubly periodic elliptic Weierstraß function

$$\wp(z) \equiv \wp(z|\omega_1, \omega_2),\qquad\text{we have } \wp(z + 2\omega_1) = \wp(z) = \wp(z + 2\omega_2)$$

and, furthermore, the theorem: If $\omega_1$ is real and $\omega_2$ purely imaginary, then $\wp(z)$ is real—which is what we want.

For us this means: Since $t = \wp(u + \beta)$ lies between $e_2$ and $e_3$ , the imaginary part of $\beta$ must be the imaginary half-period $e_2$; and $t = e_3$ when $z = 1$, then

$$\wp(\beta) = e_3 = \wp(\omega_3) \text{ and hence } \beta = \omega_3.$$

Therefore, $t = \wp(\omega + \omega_3)$. From

$$\frac{dx}{ds} = \cos\varphi = t + \frac{A}{3} = t + \frac{1}{2}e_1$$

we find

$$\frac{\sqrt{2}}{c}\frac{dx}{du} = \wp(\omega + \omega_3) + \frac{1}{2}e_1$$

or, when integrated (with Weierstraß's $\zeta$ function, not Riemann's!):

$$\wp(z) = -\zeta'(z),$$

$$\frac{\sqrt{2}}{c}x = -\zeta(u + \omega_3) + \frac{1}{2}e_1 u + \lambda.$$

The constant of integration can be determined by the condition $x = 0$ when $u = 0$ and hence $\lambda = \zeta(\omega_3)$.

We thus obtain

$$\frac{\sqrt{2}}{c}x = \frac{1}{2}e_1 u - \zeta(u + \omega_3) + \zeta(\omega_3).$$

Now we can write

$$\frac{2y^2}{c^2} = A - z = A - t - \frac{A}{3} = \frac{2}{3}A - t$$

$$= e_1 - \wp(u + \omega_3). \tag{4.10}$$

Let us assume the distance between the plates to be $2a$. Then, corresponding to $x = a$ the value of $u$ is given by

$$\sin \alpha = z = \wp(u + \omega_3) + \frac{A}{3}.$$

Again without proof, we make use of the formula

$$\wp(u + \omega_3) = e_3 + \frac{(e_3 - e_1)(e_3 - e_2)}{\wp(u) - e_3}, \tag{4.11}$$

so that we obtain

$$\sin \alpha = \frac{A}{3} + e_3 + \frac{(e_3 - e_1)(e_3 - e_2)}{\wp(u) - e_3}$$

$$= 1 + \frac{2(1 - A)}{\wp(u) - 1 + \frac{A}{3}}.$$

In fact, we derive the result

$$\wp(u) = \frac{A(5 + \sin \alpha)/3 - (1 + \sin \alpha)}{3(1 - \sin \alpha)}. \tag{4.12}$$

Finally, we can write (4.10) with the aid of (4.11) in the form

$$2\frac{y^2}{c^2} = e_1 - \wp(u + \omega_3) = e_1 - e_3 - \frac{(e_3 - e_1)(e_3 - e_2)}{\wp(u) - e_3}$$

$$= (A - 1)\left(1 + \frac{(e_3 - e_2)}{\wp(u) - e_3}\right) = (A - 1)\frac{\wp(u) - e_3 + e_3 - e_2}{\wp(u) - e_3}$$

$$= (A - 1)\frac{\wp(u) - e_2}{\wp(u) - e_3}.$$

Here, then, is our final result:

$$2\frac{y^2}{c^2} = (A - 1)\frac{\wp(u) - e_2}{\wp(u) - e_3} \,. \qquad A = 3\alpha\,, \qquad A - 1 = 3\alpha - 1$$

$$= e_1 - e_3\,, \qquad (4.13)$$

or, with

$$\kappa = \frac{4}{c^2}y \,:\, \kappa^2 = \frac{8}{c^2}\frac{\wp(u) - e_2}{\wp(u) - e_3}(A - 1) = \left(\frac{d\varphi}{ds}\right)^2 \qquad (4.14)$$

$$\kappa = \frac{2\sqrt{2}}{c}\sqrt{\frac{\wp(u) - e_2}{\wp(u) - e_3}(e_1 - e_3)} = \frac{d\varphi}{ds}\,, \qquad u = \frac{\sqrt{2}}{c}s\,, \qquad c = 2\sqrt{\frac{\gamma}{g\sigma}}\,. \quad (4.15)$$

# Chapter 5
# A Final Application in Elasticity with Jacobi Elliptic Functions

Last of all, we want to return to the *curva elastica*, but this time from the engineering point of view. This is after all a problem which arose in statics of deformable bodies, where deflection of beams is of central interest and already caught Galileo's attention as early as 1638.

In the following centuries remarkable progress was made, especially after calculus was invented, which provided the necessary tool for treating the problem analytically after a period where geometric construction pervaded much of the early work on elasticity.

It would be interesting to start a long historical discussion of how calculus allowed a more deeply mathematical understanding of, e.g., the importance of curvature. We will instead adopt a more modern treatment from the engineering perspective, making extensive use of elliptic functions of the Jacobian type.

Although the wording applied by practitioners to the realm of elasticity requires some familiarity, we will nevertheless be using their terminology.

So again, we consider the problem of determining the forms which a rod, straight and prismatic in the unstressed state, can take when held by forces and couples (i.e., pairs of equally opposite forces equally distant from the midpoint of their connecting axis) applied at the rod's extremities only, so that it is bent in a principal plane, and the central line of the rod becomes a plane curve.

The shape of the *elastica* is determined by the equation

$$\frac{B}{2}\frac{d^2\theta}{ds^2} + R\sin\theta = 0 \tag{5.1}$$

or, when multiplied by $\frac{d\theta}{ds}$:

$$\frac{B}{2}\frac{d}{ds}\left(\frac{d\theta}{ds}\right)^2 - R\frac{d}{ds}\cos\theta = 0,$$

© The Author(s), under exclusive license to Springer Nature Switzerland AG 2021
W. Dittrich, *The Development of the Action Principle*,
SpringerBriefs in Physics,
https://doi.org/10.1007/978-3-030-69105-9_5

**Fig. 5.1** Form of *elastica* in
elasticity

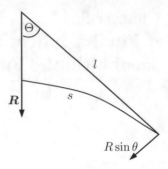

so that

$$\frac{B}{2}\left(\frac{d\theta}{ds}\right)^2 - R\cos\theta = const. \tag{5.2}$$

is our first integral of Eq. (5.1).

In Fig. 5.1, $\theta$ is the angle with which the tangent of the *lemniscate* line at any point drawn on the $s$ line increases with $s$. The line of action of the force $r$ is in the $x$ direction and is applied at the extremity from which $s$ is measured.

Finally, the stress couple is a flexural couple $B\kappa = B\frac{1}{\rho} = B\frac{d\theta}{ds}$ with a shearing force directed towards the center of curvature (*radius osculi*). We shall consider two forms of the *elastica* viz. inflexional or non-inflexional elastica according to whether there are or are not inflexions. At an inflexion, $\frac{d\theta}{ds}$ vanishes and so does the flexional couple, so that the rod can be held in the form of an inflexional *elastica* by terminal force alone, without couple. The end points are the inflexions, and it is clear that all the inflexions lie on the line of action of the terminal force $R$—the line of thrust. The kinematic analogue of an inflexional *elastica* is an oscillating pendulum.

To hold the rod with its central line in the form of a non-inflexional *elastica*, terminal couples are required as well as terminal forces. The kinematic analogue is a revolving pendulum.

## 5.1   Inflexional Elastica

If $s$ be measured from an inflexion and $\alpha$ be the value of $\theta$ at the inflexion $s = 0$, the Eq. (5.2) in this case becomes

$$\frac{B^2}{2}\left(\frac{d\theta}{ds}\right)^2 + R(\cos\alpha - \cos\theta) = 0 \tag{5.3}$$

i.e.,

$$\left(\frac{d\theta}{ds}\right)^2 = \frac{4R}{B}\left(\sin^2\frac{\alpha}{2} - \sin^2\frac{\theta}{2}\right).$$

Integrating, we obtain

$$2\sqrt{\frac{R}{B}}\int_0^s ds = \int_0^\theta \frac{d\theta}{\sqrt{\sin^2\frac{\alpha}{2} - \sin^2\frac{\theta}{2}}}.$$

Putting $\sin\frac{\theta}{2} = u\sin\frac{\alpha}{2}$ and $k = \sin\frac{\alpha}{2}$, we find

$$\sqrt{\frac{R}{B}}s = \int_0^u \frac{du}{\sqrt{(1-u^2)(1-k^2u^2)}}.$$

This yields our first Jacobian elliptic functions, $sn(z,k)$ and $dn(z,k)$, as solutions:

$$u = sn\left(\sqrt{\frac{R}{B}}s, k\right).$$

In fact,

$$\sin\frac{\theta}{2} = \sin\frac{\alpha}{2}sn\left(\sqrt{\frac{R}{B}}s\right) = k\,sn\left(\sqrt{\frac{R}{B}}s\right),$$

and

$$\cos\frac{\theta}{2} = dn\left(\sqrt{\frac{R}{B}}s\right).$$

In order to determine the exact shape of the curve, let $(x, y)$ be the coordinates of a point referred to fixed axes so that the line of thrust coincides with the axis of $x$.
   Also

$$\frac{dx}{ds} = \cos\theta, \qquad \frac{dy}{ds} = \sin\theta.$$

Thus

$$x = \int\left(2\cos^2\frac{\theta}{2} - 1\right)ds$$

$$= \int\left[2dn^2\left(\sqrt{\frac{R}{B}}s\right) - 1\right]ds$$

$$= 2\sqrt{\frac{R}{B}}E\left(\sqrt{\frac{R}{B}}s\right) - s + C,$$

where $C$ is a constant of integration. Here we used Legendre's elliptic integral of the second kind: $E(z, k) = \int_0^z dn^2(z, k)dz$.

Furthermore

$$y = \int \sin\theta \, ds$$

$$= 2 \int \sin\frac{\theta}{2} \cos\frac{\theta}{2} ds$$

$$= 2k \int sn\left(\sqrt{\frac{R}{B}}s\right) dn\left(\sqrt{\frac{R}{B}}s\right) ds$$

$$= -2k\sqrt{\frac{R}{B}} cn\left(\sqrt{\frac{R}{B}}s\right) + C'$$

$$= -2k\sqrt{\frac{R}{B}}\sqrt{(1-u^2)} + C'.$$

So

$$x = 2\sqrt{\frac{R}{B}} E\left(\sqrt{\frac{R}{B}}s\right) - s + C$$

$$y = -2k\sqrt{\frac{R}{B}} cn\left(\sqrt{\frac{R}{B}}s\right) + C', \tag{5.4}$$

where $s$ and $u$ are connected by the relation

$$u = sn\left(\sqrt{\frac{R}{B}}s\right),$$

and the constants $C$, $C'$ are chosen so that $x$, $y$ vanishes with $s$.

We also find from $\left(\frac{d\theta}{ds}\right)^2 = \frac{4R}{B}\left(\sin^2\frac{\alpha}{2} - \sin^2\frac{\theta}{2}\right)$, using $\sin\frac{\theta}{2} = ksn\sqrt{\frac{R}{B}}s$ and

$$k = \sin\frac{\alpha}{2}: \quad \left(\frac{d\theta}{ds}\right)^2 = \frac{4R}{B}\left(k^2 - k^2sn^2\sqrt{\frac{R}{B}}s\right) = \frac{4R}{B}k^2\left(1 - sn^2\sqrt{\frac{R}{B}}s\right),$$

which leads to

$$\frac{d\theta}{ds} = \kappa = 2\sqrt{\frac{R}{B}} kcn\sqrt{\frac{R}{B}}s.$$

## 5.2  Non-inflexional Elastica

In case there are no inflexions, Eq. (5.3) is changed to

$$\frac{B}{2}\left(\frac{d\theta}{ds}\right)^2 = R\cos\theta + R\left\{1 + \frac{2(1-k^2)}{k^2}\right\}, \tag{5.5}$$

where $k < 1$

$$\left(\frac{d\theta}{ds}\right)^2 = 2\left(\frac{R}{B}\right)\left\{\cos\theta + 1 + \frac{2(1-k^2)}{k^2}\right\}$$

$$= \frac{4}{k^2}\cdot\frac{R}{B}\left(1 - k^2\sin^2\frac{\theta}{2}\right).$$

Hence

$$\frac{d\theta}{ds} = \frac{2}{k}\sqrt{\frac{R}{B}}\sqrt{(1 - k^2\sin^2\frac{\theta}{2})};$$

i.e.,

$$\frac{2}{k}\sqrt{\frac{R}{B}}s = \int_0^\theta \frac{d\theta}{\sqrt{1 - k^2\sin^2\frac{\theta}{2}}}$$

$$= 2u, \qquad (\text{putting } sn\,u = \sin\frac{1}{2}\theta).$$

We thus obtain

$$u = \frac{1}{k}\sqrt{\frac{R}{B}}s.$$

Measuring $s$ from a point at which $\theta = 0$, the coordinates $(x, y)$ can be expressed as

$$x = \int_0^s \cos\theta\, ds$$

$$= \int_0^s (1 - 2\sin^2\frac{\theta}{2})ds$$

$$= k\sqrt{\frac{B}{R}}\int_0^u (1 - 2sn^2 u)du$$

$$= k\sqrt{\frac{B}{R}} \int_0^u \left\{ 1 - \frac{2(1 - dn^2 u)}{k^2} \right\} du$$

$$= k\sqrt{\frac{B}{R}} \left\{ \left( 1 - \frac{2}{k^2} \right) u + \frac{2}{k^2} E(u) \right\}.$$

Likewise,

$$y = \int_0^s \sin\theta\, ds$$

$$= 2 \int_0^s \cos\frac{\theta}{2} \sin\frac{\theta}{2}\, ds$$

$$= 2k\sqrt{\frac{B}{R}} \int_0^u cn\, u\, sn\, u\, du$$

$$= -2k\sqrt{\frac{B}{R}} \frac{dn\, u}{k^2}$$

$$= -\frac{2}{k}\sqrt{\frac{B}{R}} dn\, u.$$

Thus we obtain

$$x = k\sqrt{\frac{B}{R}} \left\{ \left( 1 - \frac{2}{k^2} \right) u + \frac{2}{k^2} E(u) \right\},$$

$$y = -\frac{2}{k}\sqrt{\frac{B}{R}} dn\, u,$$

where the constants of integration are chosen so that $x$ vanishes with $s$ and the $x$ axis is parallel to the line of action of $R$ and at such a distance from it that the force $R$ and the couple, $-B\frac{d\theta}{ds}$, which must be applied at the ends of the rod, are statically equivalent to a force $R$ acting along the axis of $x$. Actually, the curve consists of a series of loops lying altogether on one side of this axis.

# Chapter 6
# Short List of Jacobi Elliptic Functions and Constants Used in Chap. 5

The Jacobi elliptic functions $sn(z, k)$, $cn(z, k)$ and $dn(z, k)$ can be represented by means of elliptic integrals, e.g., Fig. 6.1

$$z = \int\limits_{0}^{sn(z,k)} \frac{1}{\sqrt{(1 - t^2)(1 - k^2 t^2)}} dt . \tag{6.1}$$

From here we can derive the constant $K$ as function of $k$:

$$K(k) = \int\limits_{0}^{\frac{\pi}{2}} \frac{1}{\sqrt{(1 - k^2 \sin^2 \theta)}} d\theta .$$

Proof: From (6.1) we find for $z = K$

$$K = \int\limits_{0}^{sn(K,k)} \frac{1}{\sqrt{(1 - t^2)(1 - k^2 t^2)}} dt ,$$

so that with $sn(K, k) = 1$, cf. below,

$$K = \int\limits_{0}^{1} \frac{1}{\sqrt{(1 - t^2)(1 - k^2 t^2)}} dt \quad \text{with} , t = \sin\theta$$

$$K(k) = \int\limits_{0}^{\frac{\pi}{2}} \frac{1}{\sqrt{(1 - k^2 \sin^2 \theta)}} d\theta , \quad K(0) = \frac{\pi}{2}, \quad K(1) = \infty \tag{6.2}$$

© The Author(s), under exclusive license to Springer Nature Switzerland AG 2021
W. Dittrich, *The Development of the Action Principle*,
SpringerBriefs in Physics,
https://doi.org/10.1007/978-3-030-69105-9_6

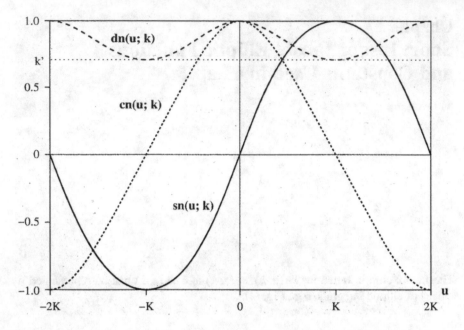

**Fig. 6.1** Elliptic functions $sn(u; 0) = \sin u$, $\quad cn(u; 0) = \cos u \quad dn(u; 0) = 1$

or

$$K(k) = F\left(\frac{\pi}{2}, k\right)$$

with

$$F(\theta, k) = \int\limits_{0}^{\theta} \frac{1}{\sqrt{(1 - k^2 \sin^2 \theta)}} d\theta,$$

the complete elliptic integral of the first kind, or

$$F(z, k) = \int\limits_{0}^{z} \frac{1}{\sqrt{(1 - t^2)(1 - k^2 t^2)}},$$

Legendre's integral of the first kind.

Likewise, the integral $\int_{0}^{z} dn^2(u, k)du$ is generally defined as Legendre's integral of the second kind and is denoted by

$$E(z, k) = \int\limits_{0}^{z} dn^2(u, k)du.$$

In particular,

$$E(K,k) = \int_0^K dn^2(u,k)du$$

and with $dn^2(z,k) + k^2 sn^2(z,k) = 1$, $am(K,k) = \frac{\pi}{2}$,    $sn(z,k) = \sin\theta$

$$E(K,k) = \int_0^{\frac{\pi}{2}} \sqrt{(1 - k^2 \sin^2\theta)}d\theta .  \tag{6.3}$$

Legendre's elliptic integral of the second kind can also be written as

$$\int \frac{z^2}{\sqrt{(1-z^2)(1-k^2z^2)}}dz ,$$

or, with $z = sn(u,k)$ this can be transformed into the form

$$\int sn^2(u,k)du$$

$$= \frac{1}{k^2} \int (1 - dn^2(u,k))du$$

$$= \frac{u}{k^2} - \frac{1}{k^2} \int (dn^2(u,k))du .$$

Since $y = snz$ satisfies a differential equation

$$\frac{dy}{dz} = \sqrt{(1-y^2)(1-k^2y^2)} ,$$

for $k \to 0$, we obtain

$$\frac{dy}{dz} = \sqrt{(1-y^2)} ,$$

so that

$$z = \int_0^y \frac{dy}{\sqrt{(1-y^2)}} = \sin^{-1} y$$

and thus, $y = \sin z$:

$$z = \int_0^{sn(z,0)} \frac{1}{\sqrt{(1-t^2)}}dt = \int_0^{\sin z} \frac{1}{\sqrt{(1-t^2)}}dt ,$$

Altogether

$$sn(z, 0) = \sin z$$
$$cn(z, 0) = \cos z$$
$$dn(z, 0) = 1 .$$

For $k \to 1$:

$$\frac{dy}{dz} = (1 - y^2)$$

$$z = \int\limits_0^y \frac{dy}{(1 - y^2)} = \tan h^{-1} y$$

$$y = \tan hz .$$

Finally,

$$sn(z, 1) = \tan hz .$$
$$cn(z, 1) = sech\, z = \frac{1}{\cos hz} .$$
$$dn(z, 1) = sech\, z .$$

Without proof, for complex values:

$$sn(K + iK'; k) = \frac{1}{k} .$$

So we can write

$$K + iK' = \int\limits_0^{sn(K+iK';k)} \frac{1}{\sqrt{(1 - t^2)(1 - k^2 t^2)}} dt$$

$$= \int\limits_0^{\frac{1}{k}} \frac{1}{\sqrt{(1 - t^2)(1 - k^2 t^2)}} dt .$$

Using the value for $K$ from (6.2), we can easily derive

$$K(k') \equiv K' = \int\limits_0^{\frac{\pi}{2}} \frac{1}{\sqrt{(1 - k'^2 \sin^2 \theta)}} d\theta , \qquad k'^2 = (1 - k^2) . \qquad (6.4)$$

Last of all we mention the periodicity of the Jacobian elliptic functions:

$$sn(z + 4K) = snz, * \qquad sn(z + 2iK') = snz, **$$
$$cn(z + 4K) = cnz, \qquad cn(z + 2K + 2iK') = cnz,$$
$$dn(z + 2K) = dnz, \qquad dn(z + 4iK') = dnz. \tag{6.5}$$

The formulae in (6.5) marked by stars are indicative of the real and imaginary frequencies of the doubly periodic mathematical pendulum.

# Chapter 7
# Variational Methods for Periodic Motions; Mathieu Functions

A new, highly interesting equation in mathematical physics appeared in 1868 when Emile Leonard Mathieu (1855–1890) studied wave equations with elliptical boundaries, i.e., when in cylindrical problems with circular boundary conditions, the latter were changed into elliptical ones. The associated equations of motion are called Mathieu's equations of motion, and their solutions are named Mathieu functions.

A prototype of Mathieu's equation is a linear differential equation with periodic coefficients and is conveniently written as (see Whittaker and Watson, Modern analysis, Chap. XIX)

$$\frac{d^2x}{dt^2}(a + 16q \cos 2t)x = 0, \quad a, q \, const. \tag{7.1}$$

Periodic solutions of this type of equation are used in solving boundary-value problems such as those of waveguides of elliptical cross-section, tides in elliptical lakes, and vibrations of loudspeakers and drumheads of elliptical shape.

Another problem whose solution requires the use of Mathieu's functions is the motion of an ideal simple pendulum with a vertically moving support; more specifically, the motion of a particle of unit mass, which is suspended by a weightless rigid rod of length $l$, which moves in the $x$–$y$ plane acted upon by gravity $g$, with the $y$-direction positive downward (Fig. 7.1).

The point of support $S$ moves vertically with varying height $y_s(t)$, regarded as an arbitrary but given function; there is no horizontal motion of $S$, i.e., $x_s(t) = 0$.

Our goal is to find the equation of motion for the angle $\theta$ using the action principle as represented in the Euler–Lagrange form.

Before we start with the detailed analytical formulation of the problem, a few words are in order as to what kind of special function to expect. The type of equation of motion we are interested in is evidently

**Fig. 7.1** Pendulum with
oscillating support

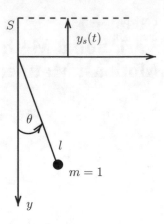

$$\ddot{x}(t) + f(t)x(t) = 0\,,$$

which can be derived from

$$\frac{d}{dt}\frac{\partial L}{\partial \dot{x}} - \frac{\partial L}{\partial x} = 0\,.$$

Defining $p_x = \dot{x}$ (with mass $m = 1$) $= \frac{\partial L}{\partial \dot{x}}$, we can write

$$L = \frac{1}{2}\dot{x}^2 + g(x,t)\,, \qquad \frac{\partial L}{\partial x} = \frac{d}{dt}\frac{\partial L}{\partial \dot{x}} = \ddot{x} = -f(t)x\,,$$

which can be satisfied by

$$L(x, \dot{x}; t) = \frac{1}{2}(\dot{x}^2 - f(t)x^2)\,.$$

Now, if we specify the function $f(t)$ by

$$f(t) = (a + 16q \cos 2t)\,,$$

we obtain the conventionally written Mathieu equation,

$$\frac{d^2x}{dt^2} + (a + 16q \cos 2t)x = 0$$

and

$$L(x, \dot{x}; t) = \frac{1}{2}\left[\dot{x}^2 - (a + 16q \cos 2t)x^2\right]\,. \tag{7.2}$$

For our specific example we want to find the equation of motion for the angle $\theta$ of inclination.

Hence, in order to apply the action principle, we need to know the Lagrangian $L(\theta, \dot{\theta}, y_s(t))$. Let us start with the conventional definition of $L$:

$$L = T - V; \qquad T = \frac{1}{2}(\dot{x}^2 + \dot{y}^2)$$

and with $m = 1$, $V = -gy$.

From Fig. 7.1 we have:

$$x = l\sin\theta \qquad y = l\cos\theta + y_s(t)$$
$$\dot{x} = \dot{\theta}l\cos\theta, \qquad \dot{y} = \dot{y}_s - \dot{\theta}l\sin\theta,$$

so that we can write

$$L(\theta, \dot{\theta}, y_s(t)) = \frac{l^2}{2}\left[\dot{\theta}^2 - 2\frac{\dot{y}_s}{l}\dot{\theta}\sin\theta + \left(\frac{\dot{y}_s}{l}\right)^2\right] + gl\left[\cos\theta + \frac{y_s(t)}{l}\right]. \qquad (7.3)$$

Given this Lagrangian, we find for the Euler–Lagrange equations using:

$$\frac{\partial L}{\partial \dot{\theta}} = l^2\left(\dot{\theta} - \frac{\dot{y}_s}{l}\sin\theta\right), \qquad \frac{\partial L}{\partial \theta} = -l\dot{y}_s\dot{\theta}\cos\theta - gl\sin\theta,$$

$$\frac{d}{dt}\left(\frac{\partial L}{\partial \dot{\theta}}\right) = l^2\left[\ddot{\theta} - \frac{\ddot{y}_s}{l}\sin\theta - \frac{\dot{y}_s\dot{\theta}}{l}\cos\theta\right] = \frac{\partial L}{\partial \theta} = -l\dot{y}_s\dot{\theta}\cos\theta - gl\sin\theta,$$

$$(7.4)$$

from which we obtain the sought-for equation of motion for the angle $\theta(t)$:

$$\ddot{\theta} + \frac{1}{l}[g - \ddot{y}_s(t)]\sin\theta = 0. \qquad (7.5)$$

Incidentally, this equation of motion could have been obtained directly from Einstein's equivalence principle (general relativity), which states that a gravitational field g is equivalent in all respects to an acceleration $a$ in the opposite direction; $a = -g$. Consequently, the acceleration $\ddot{y}_s$ is equivalent to an opposite contribution to gravity: $g_s = -\ddot{y}_s$. So the total gravitation is indeed given by $g_{tot} = g - \ddot{y}_s$.

In the following we want to take a closer look at a small-amplitude pendulum. For this case, it is easy to find the linearized form of the equations of motion and the corresponding Lagrangian, all for an arbitrary function $y_s(t)$. So we approximate $\sin\theta$ by $\theta$ and obtain the following equations:

$$\ddot{\theta} + \frac{1}{l}[g - \ddot{y}_s]\theta = 0\,,$$

$$\frac{\partial L}{\partial \dot{\theta}} = l^2\dot{\theta}\,, \qquad \frac{d}{dt}\left(\frac{\partial L}{\partial \dot{\theta}}\right) = l^2\ddot{\theta} : L = \frac{l^2}{2}\dot{\theta}^2 \qquad (7.6)$$

$$\frac{\partial L}{\partial \theta} = l^2\ddot{\theta} = -l[g - \ddot{y}_s(t)]\theta\,. \qquad (7.7)$$

These equations can be satisfied by

$$L \cong \frac{l^2}{2}\left[\dot{\theta}^2 - \frac{1}{l}(g - \ddot{y}_s)\theta^2\right]\,. \qquad (7.8)$$

Now let us assume that $y_s(t)$ is harmonic. Then all the former functions and equations in the small-amplitude limit will take on the following form:

$$\text{for} \quad y_s(t) = -A\cos\omega t\,, \qquad \ddot{y}_s = \omega^2 A\cos\omega t \qquad (7.9)$$

we obtain

$$L = \frac{l^2}{2}\left[\dot{\theta}^2 - \frac{1}{l}(g - \omega^2 A\cos\omega t)\theta^2\right] \qquad (7.10)$$

and the equation of motion (7.5) becomes for a small angle $\theta(t)$:

$$\ddot{\theta}(t) + \frac{1}{l}(g - \omega^2 A\cos\omega t)\theta(t) = 0\,. \qquad (7.11)$$

This is indeed Mathieu's equation (7.1) if we introduce the following substitutions:

$$\theta \to x\,, \qquad \omega t \to 2\tau + \pi\,, \qquad \frac{d}{dt} = \frac{\omega}{2}\frac{d}{d\tau}\,,$$

$$\frac{4g}{\omega^2 l} = a\,, \qquad \frac{4A}{l} = 16q\,. \qquad (7.12)$$

With these changes, we finally end up with

$$\frac{d^2 x}{d\tau^2} + (a + 16q\cos 2\tau)x = 0\,.$$

After these preliminaries we want to show that the solutions of Mathieu's equation are characterized by zones of stability in the $a$-$q$ parameter plane. On the lines of this plane which separate stable from unstable regions, the solutions are periodic. For $q = 0, a > 0$, the solutions are obviously stable. For $a > 0$ and $q$ very small (near the positive a-axis), narrow zones of instability touch the $a$-axis at points where $a$ has certain simple values.

In order to solve this stability problem, we start with the principle of stationary action (Euler–Lagrange),

$$\delta S = \delta \int_{t_1}^{t_2} dt\, L = 0\,,$$

where the action functional $S$ is given by

$$S\{[x(t)]\,;\, t_1, t_2\} = \int_{t_1}^{t_2} dt\, L(x(t), \dot{x}(t); t)\,.$$

In our specific case, we are given (replace $\left(\frac{2}{l^2}L\right)$ by $L$ in (7.10)

$$L = \dot{\theta}^2 - \frac{g}{l}\theta^2 + \left(\frac{\omega^2 A}{l}\cos\omega t\right)\theta^2\,,$$

with trial function

$$\theta(t) = B\sin\frac{\omega t}{2}\,,$$

$$\dot{\theta}(t) = \frac{\omega B}{2}\cos\omega t \tag{7.13}$$

and we are looking for a stationary value for $B \neq 0$.

Therefore we require that the time integral over one period $\left(\Delta t = \frac{2\pi}{\omega}\right)$ have a stationary value for non-zero $B$:

$$\int_0^{2\pi/\omega} L(\theta, \dot{\theta})dt = \left(\frac{\omega B}{2}\right)^2 \int \cos^2\frac{\omega t}{2}dt - \frac{g}{l}B^2\int \sin^2\frac{\omega t}{2}dt$$

$$+ \frac{\omega^2 A B^2}{l}\int \cos\omega t\left(\frac{1-\cos\omega t}{2}\right)dt$$

$$\propto \left[\left(\frac{\omega}{2}\right)^2 - \frac{g}{l} - \frac{1}{2}\frac{\omega^2 A}{l}\right]B^2 = [\%]B^2\,.$$

For an extremum we require $\frac{\partial}{\partial B}[\%]B^2 = 0$, for $B \neq 0$, so that $[\%] = 0$, or

$$\left(\frac{\omega}{2}\right)^2 = \frac{g}{l} \pm \frac{\omega^2 A}{2l}\,. \tag{7.14}$$

Both signs in (7.14) appear since the origin of time is irrelevant: only the sign of $A$ is changed. Formula (7.14) is the result of the underlined assumption (trial function) that $\theta$ is a simple periodic function with period twice that of $y_s$.

**Fig. 7.2** Stability plot

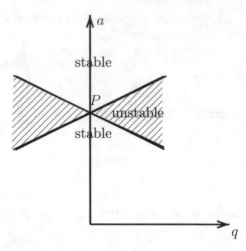

Because $\left|\frac{A}{l}\right| \ll 1$, we are very near the $q$-axis ($|q| = \left|\frac{A}{4l}\right|$) in the $a$-$q$ plane. Thus the trial function is valid very near $(\frac{\omega}{2})^2 = \frac{g}{l}$, which is the same as near $a = 1$. The stability plot Fig. 7.2 illustrates (roughly) our findings for small $|q|$ in the $a$-$q$ plane.

The straight lines crossing at $P(a = 1, q = 0)$ represent the periodic solutions which are given by

$$\left(\frac{\omega}{2}\right)^2 \simeq \frac{g}{l} \pm \frac{\omega^2 A}{2l},$$

$$\text{or} \quad \left(\frac{2}{\omega}\right)^2 \frac{g}{l} \simeq 1 \pm \frac{2A}{l},$$

$$\text{or} \quad a \simeq 1 \pm 8q. \tag{7.15}$$

Finally we want to find out how sensitive our result (7.14) is to changes in the functional form of the trial function. In order to give a little twist to our calculations, we will use the principle of stationary action as realized by the Hamiltonian version: $\delta \int (p\dot{q} - H)dt = 0$. For our specific problem, this means computing the variation

$$\delta \int (p\dot{\theta} - H(\theta, p))dt = \delta \int \left[ p\dot{\theta} - \frac{p^2}{2} - \frac{1}{2l}(g - \ddot{y}_s)\theta^2 \right] dt = 0.$$

We want to make two separate calculations, each with the trial function $\theta(t) = B \sin \frac{\omega t}{2}$. For the first one, we choose the trial function $p_1 = C_1 \cos \frac{\omega t}{2}$; for the second, we use $p_2 = C_2 \left[ \left( \frac{\pi}{\omega} \right) - t \right]$. In each calculation, we require that the integral over one period have a stationary value with regard to the parameters $B$ and $C$.

Here is the first calculation with $\theta(t) = B \sin \frac{\omega t}{2}$, $p = C_1 \cos \frac{\omega t}{2}$ and $\ddot{y}_s = \omega^2 A \cos \omega t$. Then we need

$$\int\limits_{0}^{2\pi/\omega} p\dot{\theta}\,dt = \frac{\omega C_1 B}{2}\int \cos^2\frac{\omega t}{2}\,dt\,,$$

$$\int \frac{p^2}{2}\,dt = \frac{C_1^2}{2}\int \cos^2\frac{\omega t}{2}\,dt\,,$$

$$\int \theta^2\,dt = B^2\int \sin^2\frac{\omega t}{2}\,dt\,,$$

$$\int \ddot{y}_s\theta^2\,dt = \omega^2 A B^2\int \cos\omega t\left(\frac{1-\cos\omega t}{2}\right)dt\,.$$

The result turns out to be

$$\int [\ ]dt \propto \left[\frac{\omega C_1 B}{2}-\frac{C_1^2}{2}-\frac{gB^2}{2l}+\frac{\omega^2 A B^2}{4l}\right] =: [\%]\,.$$

We require $\frac{\partial}{\partial B}[\%] = 0$ and $\frac{\partial}{\partial C_1}[\%] = 0$, which yields

$$\frac{\omega}{2}C_1 - \left(\frac{g}{l}\pm\frac{\omega^2 A}{2l}\right)B = 0 \quad\text{and}\quad \frac{\omega}{2}B - C_1 = 0\,.$$

So we need the solution of two simultaneous linear homogeneous equations for $B$ and $C_1$. For the solution to exist, it is necessary that the determinant of the coefficients be zero, i.e.,

$$\begin{vmatrix} \frac{\omega}{2} & -\left(\frac{g}{l}\pm\frac{\omega^2 A}{2l}\right) \\ -1 & \frac{\omega}{2} \end{vmatrix} = 0\,,$$

from which we obtain $\left(\frac{\omega}{2}\right)^2 = \frac{g}{l}\pm\frac{\omega^2 A}{2l}$, exactly as before.

The second calculation uses $\theta,\dot{\theta},\ddot{y}_s$, also as before, but $p = C_2\left(\frac{\pi}{\omega}-t\right)$. Again, we need the value of the following integrals over one period $\frac{2\pi}{\omega}$:

$$\int\limits_{0}^{2\pi/\omega} p\dot{\theta}\,dt = \frac{BC_2}{2}\int (\pi-\omega t)\cos\frac{\omega t}{2}\,dt = \frac{4BC_2}{\omega}$$

$$\int \frac{1}{2}p^2\,dt = \frac{C_2^2}{2}\left(\frac{\pi}{\omega}\right)^2\int\left[1-\frac{\omega t}{\pi}\right]^2 dt = \left(\frac{\pi}{\omega}\right)^3\frac{C_2^2}{3}\,,$$

$$\int \theta^2\,dt = \frac{\pi B^2}{\omega}\,,\quad \int \ddot{y}_s\theta^2\,dt = \omega^2 A B^2\frac{\pi}{4}\,.$$

The total result is

$$\int_0^{2\pi/\omega} [\,]dt = \left[ \frac{4}{\omega}BC_2 - \frac{1}{3}\left(\frac{\pi}{\omega}\right)^3 C_2^2 - \frac{\pi}{2\omega}\frac{g}{l}B^2 + \frac{\pi}{4\omega}\frac{\omega^2 A}{l}B^2 \right] = [\%]\,.$$

As before, requiring $\frac{\partial}{\partial B}[\%] = 0$ and $\frac{\partial}{\partial C_2}[\%] = 0$ yields

$$\frac{4}{\omega}C_2 - \frac{\pi}{\omega}\left(\frac{g}{l} \pm \frac{\omega^2 A}{2l}\right)B = 0\,, \qquad \frac{4}{\omega}B - \frac{2}{3}\left(\frac{\pi}{\omega}\right)^3 C_2 = 0\,.$$

The determinant of the coefficients needs to be zero:

$$\begin{vmatrix} \frac{4}{\omega} & -\frac{\pi}{\omega}\left(\frac{g}{l} \pm \frac{\omega^2 A}{2l}\right) \\ -\frac{2}{3}\left(\frac{\pi}{\omega}\right)^3 & \frac{4}{\omega} \end{vmatrix} = 0\,, \qquad \text{i.e.,} \qquad \left(\frac{4}{\omega}\right)^2 = \frac{2}{3}\left(\frac{\pi}{\omega}\right)^4\left(\frac{g}{l} \pm \frac{\omega^2 A}{2l}\right)\,.$$

Finally we obtain

$$\left(\frac{\omega}{2}\right)^2 = \left(\frac{\pi^4}{96}\right)\left(\frac{g}{l} \pm \frac{\omega^2 A}{2l}\right)\,,$$

exactly as before, except for the factor $\frac{\pi^4}{96}$, but $\frac{\pi^4}{96} = 1.01468\ldots$, so that the difference is $<1\frac{1}{2}\%$, which means that the result is very insensitive to moderate differences between the trial functions Fig. 7.3.

To investigate the stability of the equilibrium solution $\theta = 0$ we linearized the equation of motion (7.5) to derive the Mathieu equation (7.11), i.e.,

$$\ddot{\theta}(t) + \left(\frac{g}{l} - \frac{A}{l}\omega^2 \cos\omega t\right)\theta(t) \tag{7.16}$$

The vertically downward equilibrium at $\theta = 0$ will be stable or unstable or show a constant periodic oscillation, depending on the values $\frac{g}{l}$ and $\frac{A}{l}\omega^2$. It follows that $\theta = 0$ is stable for $\omega = 0$ (harmonic oscillator). The vertical oscillation (with $\omega \neq 0$) means that we are moving the support harmonically up and down and in this way put the pendulum in oscillation.

Remarkably, there is another equilibrium position at $\theta = \pi$ of the pendulum.

**Fig. 7.3** Trial functions $p_1$ and $p_2$

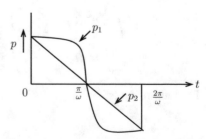

To study the stability of the vertically upward equilibrium, we introduce a new angle variable,

$$\varphi = \pi - \theta.$$

Then Eq. (7.5) turns into

$$\ddot{\varphi}(t) - \left( \frac{g}{l} - \frac{A}{l} \omega^2 \cos \omega t \right) \sin \varphi = 0. \tag{7.17}$$

When we linearize this equation around $\varphi = 0$, we find a similar Mathieu equation:

$$\ddot{\varphi}(t) - \left( \frac{g}{l} - \frac{A}{l} \omega^2 \cos \omega t \right) \varphi(t) = 0. \tag{7.18}$$

The difference of this equation to (7.16) is that the new equation of motion is unstable for $\omega = 0$. However, the vertical oscillation of the support is a means to make the pendulum motion stable. This means (surprisingly ?!) that we are able to keep the pendulum almost upward by harmonically moving the support up and down.

# Chapter 8
# Lagrangian for Isentropic Irrotational Flow

Since we are interested in formulating an action principle with an action function for fluid dynamics—at least an approximation thereof—which yields the Lagrange equations of motion, we need an appropriate Lagrange function with its relevant variables to apply the necessary variational calculations. This will *not* be the velocity $\vec{v}(t)$ in (8.1). Things become even more complicated when trying to extend the search for a Lagrangian and its dynamical variables that formulate those Lagrange equations of motion and their solutions that describe the behavior of a real fluid (Navier–Stokes). The interested reader can find more details on the subject in [1].

Here it might help to think of classical electrodynamics: In chapter 9 we will find the necessary exact Lagrangian and the set of dynamical variables to be varied in order to obtain the Lagrange equations of motion [2]. These variables will *not only* be the electromagnetic field $\vec{E}(\vec{r}, t)$ and $\vec{B}(\vec{r}, t)$ which are the building blocks of Maxwell's equations!

The starting point for this chapter in which we study the dynamics of fluids is Euler's equation of motion for the velocity field $\vec{v}(\vec{r}, t)$ in nonviscous hydrodynamics:

$$\frac{d\vec{v}}{dt} \equiv \frac{\partial \vec{v}}{\partial t} + \left( \vec{v} \cdot \vec{\nabla} \right) \vec{v} = -\frac{1}{\varrho} \vec{\nabla} p + \vec{f}. \tag{8.1}$$

Note the intrinsic nonlinearity which accounts for the difficulty of hydrodynamics. Equation (8.1) describes, by the way, the first classical (*non-linear*) field theory in history. It is the velocity field vector $\vec{v}(\vec{r}, t)$ in a surrounding fluid and not the electromagnetic field vectors $\vec{E}(\vec{r}, t)$ and $\vec{B}(\vec{r}, t)$ of classical electrodynamics, which is a *linear* field theory.

To continue, we transform Eq. (8.1) further with the aid of the vector identity

$$(\vec{v} \cdot \vec{\nabla}) \vec{v} \equiv \vec{\nabla} \left( \frac{1}{2} v^2 \right) - \left[ \vec{v} \times \left( \vec{\nabla} \times \vec{v} \right) \right],$$

which leads to

$$\frac{\partial \vec{v}}{\partial t} + \vec{\nabla}\left(\frac{1}{2}v^2\right) - \vec{v} \times \left(\vec{\nabla} \times \vec{v}\right) = \vec{f}_{\text{app}} - \frac{1}{\varrho}\vec{\nabla}p, \tag{8.2}$$

where $p$ is the exerted pressure on the volume of the fluid, $\vec{f}_{\text{app}}$ is an applied force per unit mass of fluid, and $\varrho$ is the mass density of the fluid.

In addition, there is also for a given element of fluid with total mass $M$ an internal energy $M\varepsilon$ occupying a volume $V$. Now, for any reversible isentropic ($\Delta S = 0$) process, the first law of thermodynamics relates the increase in the internal energy to the work done on the element, as seen by the comoving observer.

Altogether there are three sources that contribute to the increase of the total energy in $V$ : convection of kinetic and internal energy, work done by the pressure force on the surface, and work done by the volume force $\vec{f}$. Then the result for the energy conservation law is contained in the differential form:

$$\frac{\partial}{\partial t}\left(\frac{1}{2}\varrho v^2 + \varrho \varepsilon\right) + \vec{\nabla} \cdot \left[\left(\frac{1}{2}\varrho v^2 + \varrho \varepsilon + p\right)\vec{v}\right] = \varrho \vec{f} \cdot \vec{v}, \tag{8.3}$$

which identifies the energy flux vector,

$$\vec{j}_e = \left(\frac{1}{2}\varrho v^2 + \varrho \varepsilon + p\right)\vec{v}. \tag{8.4}$$

There is, of course, also the differential continuity equation, which stands for conservation of (fluid) matter:

$$\frac{\partial \varrho}{\partial t} + \vec{\nabla} \cdot (\varrho \vec{v}) = 0; \tag{8.5}$$

in words: the time derivative of the mass density is related to the corresponding mass current density.

Expanding the divergence of $\varrho \vec{v}$, we can also rewrite the continuity equation as

$$\frac{\partial \varrho}{\partial t} + \vec{v} \cdot \vec{\nabla}\varrho + \varrho \vec{\nabla} \cdot \vec{v} = 0, \tag{8.6}$$

or, using

$$\frac{d\varrho}{dt} \equiv \frac{\partial \varrho}{\partial t} + \vec{v} \cdot \vec{\nabla}\varrho, \tag{8.7}$$

$$\frac{d\varrho}{dt} + \varrho \vec{\nabla} \cdot \vec{v} = 0. \tag{8.8}$$

In the particular case of a uniform incompressible fluid, $\frac{\partial \varrho}{\partial t}$ and $\frac{d\varrho}{dt}$ vanish identically. So we obtain

$$\vec{\nabla} \cdot \vec{v} = 0 \tag{8.9}$$

for a divergenceless velocity for an incompressible flow.

Despite the nonlinearity of Euler's Eq. (8.1), there exists an exact first integral in the important case of irrotational flow $\left(\vec{\nabla} \times \vec{v}\right) = 0$. To prove this, let us go back to Eq. (8.2) and impose the following restrictions:

1.   The motion is irrotational:

$$\vec{\nabla} \times \vec{v} = 0. \tag{8.10}$$

Consequently, the velocity $\vec{v}$ can be derived from a scalar velocity potential $\Phi$, according to.

$$\vec{v} = -\vec{\nabla}\Phi. \tag{8.11}$$

2.   The applied force is conservative, so that

$$\vec{f} = -\vec{\nabla}U(\vec{r}, t), \tag{8.12}$$

where $U$ is the potential energy per unit mass. Here, the external potential is still permitted to have an arbitrary time dependence.

3.   The fluid is incompressible with fixed constant density:

$$\varrho = const. \tag{8.13}$$

Under these three restrictions, we are allowed to rewrite equation

$$\vec{\nabla}\left(\frac{p}{\varrho} + U + \frac{1}{2}v^2 - \frac{\partial \Phi}{\partial t}\right) = 0, \tag{8.14}$$

which can be integrated with the result

$$\frac{p}{\varrho} + U + \frac{1}{2}v^2 - \frac{\partial \Phi}{\partial t} = \Lambda(t), \tag{8.15}$$

where $\Lambda(t)$ is solely a function of time.

Equation (8.15) is reminiscent of a gauge transformation in electrodynamics. In fact, we can gauge-transform the "old" velocity potential $\overline{\Phi}$ into a new velocity potential $\Phi$ to make $\Lambda(t)$ vanish:

$$\Phi(\vec{r}, t) = \overline{\Phi}(\vec{r}, t) + \int^{t} dt' \Lambda(t'). \tag{8.16}$$

Evidently, by taking the gradient $\vec{\nabla}$ on both sides of (8.16), we obtain the same velocity. Hence, when putting $\Lambda(t) = 0$ in (8.15), we obtain the (Bernoulli) theorem,

$$\frac{p}{\varrho} + U + \frac{1}{2}v^2 - \frac{\partial \Phi}{\partial t} = 0. \tag{8.17}$$

This equation holds for an incompressible irrotational flow. But the restriction to an incompressible fluid is readily relaxed to include compressible isentropic flow as considered in (8.4), so that Bernoulli's theorem for isentropic irrotational flow has the more general form:

$$\varepsilon + \frac{p}{\varrho} + U + \frac{1}{2}v^2 - \frac{\partial \Phi}{\partial t} = 0. \tag{8.18}$$

Together with the continuity equation,

$$\frac{\partial \varrho}{\partial t} - \vec{\nabla} \cdot \left(\varrho \vec{\nabla} \Phi\right) = 0 \tag{8.19}$$

we want to show that these two equations are precisely the Euler–Lagrange equations of the Lagrangian density with generalized fields $\Phi$ and $\varrho$ :

$$\boxed{\mathcal{L}(\Phi, \varrho) = \varrho \frac{\partial \Phi}{\partial t} - \frac{1}{2}\varrho\left(\vec{\nabla}\Phi\right)^2 - \varrho U - \varrho \varepsilon(\varrho).} \tag{8.20}$$

From this expression we obtain the canonical momentum density,

$$\Pi_{\Phi} = \frac{\partial \mathcal{L}}{\partial\left(\frac{\partial \Phi}{\partial t}\right)} = \varrho, \tag{8.21}$$

which obeys the (Lagrange) equation of motion,

$$\frac{\partial}{\partial t}\Pi_{\Phi} + \vec{\nabla} \cdot \frac{\partial \mathcal{L}}{\partial\left(\vec{\nabla}\Phi\right)} = \frac{\partial \mathcal{L}}{\partial \Phi}, \tag{8.22}$$

which is just the continuity Eq. (8.19).

The second canonical momentum density $\Pi_{\varrho}$ vanishes identically, since $\mathcal{L}$ does not contain $\frac{\partial \varrho}{\partial t}$.

$$\Pi_\varrho = \frac{\partial \mathcal{L}}{\partial \left( \frac{\partial \varrho}{\partial t} \right)} = 0. \tag{8.23}$$

Finally, the corresponding dynamical Lagrange equation,

$$\frac{\partial}{\partial t} \frac{\partial \mathcal{L}}{\partial \left( \frac{\partial \varrho}{\partial t} \right)} + \vec{\nabla} \cdot \frac{\partial \mathcal{L}}{\partial \left( \vec{\nabla} \varrho \right)} = \frac{\partial \mathcal{L}}{\partial \varrho}, \tag{8.24}$$

is just Bernoulli's Eq. (8.18), because $\frac{\partial (\varrho \varepsilon)}{\partial \varrho} = \varepsilon + \varrho^{-1} p$, which follows from $V = \frac{M}{\varrho}$ and $M d\varepsilon = -p dV = \frac{Mp}{\varrho^2} d\varrho$.

So we achieved our goal, namely, under the restrictions 1 to 3 to derive a Lagrangian which provided us with the equations of motion for the generalized fields $\Phi$ and $\varrho$.

If the applied force is time independent so that $U(\vec{r}, t) \equiv U(\vec{r})$, the absence of explicit time dependence in $\mathcal{L}$ implies that the volume integral of the Hamiltonian density

$$\mathcal{H} = \Pi_\Phi \frac{\partial \Phi}{\partial t} + \Pi_\varrho \frac{\partial \varrho}{\partial t} - \mathcal{L} = \frac{1}{2} \varrho \left( \vec{\nabla} \Phi \right)^2 + \varrho U + \varrho \varepsilon(\varrho) \tag{8.25}$$

is a constant of the motion. It is clearly the energy density, for it contains the kinetic, potential and internal energy density. As in any field theory, the Lagrangian (density) also provides a derivation of the energy flux vector $\vec{j}_e$.

With the two generalized coordinates $\varrho$ and $\Phi$, we immediately obtain

$$\vec{j}_e = \frac{\partial \mathcal{L}}{\partial \left( \vec{\nabla} \Phi \right)} \frac{\partial \Phi}{\partial t} + \frac{\partial \mathcal{L}}{\partial \left( \vec{\nabla} \varrho \right)} \frac{\partial \varrho}{\partial t} = -\varrho \vec{\nabla} \Phi \frac{\partial \Phi}{\partial t}$$

$$= -\varrho \vec{\nabla} \Phi \left[ \varepsilon + \varrho^{-1} p + U + \frac{1}{2} \left( \vec{\nabla} \Phi \right)^2 \right], \tag{8.26}$$

where in the last line we applied Bernoulli's theorem (8.18).

The general equation of motion of hydrodynamics is the Navier–Stokes equation. It is the generalization of Euler's equation for viscous fluids and is given by

$$\frac{\partial \vec{v}}{\partial t} + (\vec{v} \cdot \vec{\nabla}) \vec{v} = -\frac{1}{\varrho} \vec{\nabla} p + \vec{f} + \frac{\eta}{\varrho} \vec{\nabla}^2 \vec{v} + (\eta + \eta') \vec{\nabla} (\vec{\nabla} \cdot \vec{v}). \tag{8.27}$$

The integration of this equation is very hard indeed; this is also true for finding a corresponding Lagrangian. No wonder that the search for a solution of the Navier–Stokes equation is on the one-million dollar list of unsolved problems!

Nevertheless, for incompressible fluids, $\left(\vec{\nabla} \cdot \vec{v}\right) = 0$, there are useful solvable problems. Here is one: the stationary laminar flow of an incompressible fluid through a cylindrical pipe. To deal with this problem, we need the modified velocity field equation,

$$\frac{\partial \vec{v}}{\partial t} + (\vec{v} \cdot \vec{\nabla})\vec{v} = \vec{f} - \frac{1}{\varrho}\vec{\nabla}p + \frac{\eta}{\varrho}\vec{\nabla}^2\vec{v}. \tag{8.28}$$

With regard to a Cartesian coordinate system, we have $\vec{r} = \{x, y, z\}$,

$$\vec{v} = \{u, v, w\}, \vec{\nabla} = \left\{\frac{\partial}{\partial x}, \frac{\partial}{\partial y}, \frac{\partial}{\partial z}\right\}$$

The axis of the cylindrical pipe is in z-direction; the radius is denoted by $a$; the length by $l$. The pipe cross section lies in the x–y plane. The pressure is given by $p(z = 0) = p_1, p(z = l) = p_2$, with $p_2 < p_1$; $p(\vec{r}) = p(0, 0, z) \equiv p(z)$ is independent of $x, y$.

There are neither x- nor y-components of $\vec{v}$ : $u = 0 = v$. Since we are considering an incompressible fluid, we have.

$$\vec{\nabla} \cdot \vec{v} = 0 \quad \text{or} \quad \frac{\partial u}{\partial x} + \frac{\partial v}{\partial y} + \frac{\partial w}{\partial z} = 0, \quad \text{i.e.,} \frac{\partial w}{\partial z} = 0. \tag{8.29}$$

Under these conditions, the equation of motion (8.28) is reduced to (stationary $\frac{\partial \vec{v}}{\partial t} = 0, \vec{f} = 0$)

$$\frac{\partial p}{\partial x} = 0, \frac{\partial p}{\partial y} = 0, \vec{\nabla}^2\vec{v} = \frac{\partial^2 w}{\partial x^2} + \frac{\partial^2 w}{\partial y^2} \tag{8.30}$$

$$\frac{\partial p}{\partial z} = \eta\left(\frac{\partial^2 w}{\partial x^2} + \frac{\partial^2 w}{\partial y^2}\right) \tag{8.31}$$

In these equations, the laminar current is represented by a pure viscous phenomenon.

According to (8.29), the velocity $w$ depends only on $x$ and $y$, i.e., for a circular cross section on $r = \sqrt{x^2 + y^2}$. Likewise, according to (8.30), the pressure $p$ depends only on $z$, the direction of the cylindrical pipe.

Hence, Eq. (8.31) takes on the simple form (remember in cylindrical coordinates $\vec{\nabla}^2\varphi = \frac{1}{r}\frac{\partial}{\partial r}\left(r\frac{\partial\varphi}{\partial r)} + \ldots\right)$:

$$\frac{dp(z)}{dz} = \eta\frac{1}{r}\frac{d}{dr}\left(r\frac{dw(r)}{dr}\right). \tag{8.32}$$

Since the left-hand side is only a function of $z$, while the right-hand side depends only on $r$, both sides have to be a constant, say $-C$. Hence, we have to solve the two elementary equations

$$\frac{dp}{dz} = -C \tag{8.33}$$

and

$$\frac{dp}{dr}\left(r\frac{dw}{dr}\right) = -\frac{C}{\eta}r. \tag{8.34}$$

According to (8.33), the pressure with direction of the current flow decreases linearly where the constant $C$ determines the pressure drop:

$$C = \frac{p_1 - p_2}{l}.$$

Equation (8.34) needs two integrations:

$$r\frac{dw}{dr} = -\frac{C}{2\eta}r^2 + C_1$$

$$w = -\frac{C}{4\eta}r^2 + C_1 \log r + C_2.$$

The constant $C_1$ has to be zero, $C_1 = 0$, since otherwise along the pipe axis the velocity would be infinite.

The value for $C_1$ follows from the boundary condition where the velocity is zero, i.e., $w(r = 0) = 0$.

So we obtain for $C_2$ : $C_2 = \frac{Ca^2}{4\eta}$.

Finally, we find for the current velocity $w = \frac{C}{4\eta}\left(a^2 - r^2\right)$, or

$$w = \frac{p_1 - p_2}{4\eta l}\left(a^2 - r^2\right). \tag{8.35}$$

This result means that starting with the middle of the pipe ($r = 0$), the fluid velocity decreases in the form of a rotational paraboloid.

The result (8.35) provides us with the fluid volume $Q$ that flows per unit of time through a cross section of the pipe. To obtain this current strength, we only have to integrate over the pipe's cross section:

$$Q = \int w \, df = \frac{p_1 - p_2}{4\eta l} \int_0^a (a^2 - r^2) 2\pi r \, dr.$$

The integration is trivial and results in

$$Q = \frac{\pi a^4}{8\eta l}(p_1 - p_2), \tag{8.36}$$

which is known as the Hagen-Poiseuille law (1839–40). This, by the way, provides us with a means to measure the viscosity $\eta$ of a given fluid.

## References

1. A.L. Fetter, J.D. Walecka: Theoretical Mechanics of Particles and Continua, McGraw-Hill Book Company, 1980
2. Julian Schwinger, K.A. Milton, L.L. DeRaad, W-Y.Tsai: Classical Electrodynamics, Westview Press, Cambridge, Ma., USA, 1998

# Chapter 9
# Action Principle in Classical Electrodynamics

In our next example we want to develop an action (Lagrangian) formulation for electrodynamics. But before we do so, we find it helpful to discuss a similar, easier problem, namely the Lagrangian formulation of classical Newtonian mechanics. (See Chap. 2 of [1].) Here we will be using $\vec{r}$, $\vec{p}$ and the velocity $\vec{v}$ as independent coordinates. However, we will see immediately that $\vec{v}$ does not satisfy an equation of motion, i.e., $\frac{d\vec{v}}{dt}$ does not occur. Therefore $\vec{v}$ is not a dynamical variable. So let us write down a Lagrangian and study its consequences:

$$L = \vec{p} \cdot \left( \frac{d\vec{r}}{dt} - \vec{v} \right) + \frac{1}{2}mv^2 - V(\vec{r}) \qquad (9.1)$$

$$= \vec{p} \cdot \frac{d\vec{r}}{dt} - H(\vec{r}, \vec{p}, \vec{v}), \quad H = \vec{p} \cdot \vec{v} - \frac{1}{2}mv^2 + V(\vec{r}). \qquad (9.2)$$

Now we use the most general formulation of the action principle:

$$\delta W_{12} = G(t_1) - G(t_2), \qquad (9.3)$$

where

$$W_{12} = \int_{t_1}^{t_2} dt\, L\, dt. \qquad (9.4)$$

The fixed values $G_1$ and $G_2$ depend only on the endpoint path variables at the respective terminal times.

© The Author(s), under exclusive license to Springer Nature Switzerland AG 2021
W. Dittrich, *The Development of the Action Principle*,
SpringerBriefs in Physics,
https://doi.org/10.1007/978-3-030-69105-9_9

Therefore for the variation of the action we need

$$
\delta W_{12} = \int\limits_{1}^{2} dt \left\{ \vec{p} \cdot \frac{d}{dt} \delta \vec{r} - \delta \vec{r} \cdot \vec{\nabla} V + \delta \vec{p} \cdot \left( \frac{d\vec{r}}{dt} - \vec{v} \right) + \delta \vec{v} \cdot (-\vec{p} + m\vec{v}) - \frac{d(\delta t)}{dt} \left( \vec{p} \cdot \vec{v} - \frac{1}{2} m v^2 + V \right) \right\}
$$

$$
= \int\limits_{1}^{2} dt \left\{ \frac{d}{dt} [\vec{p} \cdot \delta \vec{r} - H \delta t] - \delta \vec{r} \cdot \left[ \frac{d\vec{p}}{dt} + \vec{\nabla} V \right] + \delta \vec{p} \cdot \left[ \frac{d\vec{r}}{dt} - \vec{v} \right] + \delta \vec{v} \cdot (-\vec{p} + m\vec{v}) + \delta t \frac{dH}{dt} \right\}.
$$

Then the action principle yields

$$
\delta \vec{r} : \frac{d\vec{p}}{dt} = -\frac{\partial H}{\partial \vec{r}} = -\vec{\nabla} V
$$

$$
\delta \vec{p} : \frac{d\vec{r}}{dt} = \frac{\partial H}{\partial \vec{p}} = \vec{v}
$$

$$
\delta \vec{v} : 0 = \vec{p} - \frac{\partial H}{\partial \vec{v}} = \vec{p} - m\vec{v}
$$

$$
\delta t : \frac{dH}{dt} = 0, \; H = \vec{p} \cdot \vec{v} - \frac{1}{2} m v^2 + V
$$

surface term $G = \vec{p} \cdot \delta \vec{r} - H \delta t$.

There is no equation of motion for $\vec{v}$, i.e., no $\frac{d\vec{v}}{dt}$.

Now we will repeat the former variational calculation for electrodynamics and show the following Lagrangian yields the correct equations of motion. According to [2], we begin with

$$
L = \sum_{k=1}^{n} \left[ \vec{p}_k \cdot \left( \frac{d\vec{r}_k}{dt} - \vec{v}_k \right) + \frac{1}{2} m_k v_k^2 - e_k \Phi(\vec{r}_k, t) + \frac{e_k}{c} \vec{v}_k \cdot \vec{A}_k(\vec{r}_k, t) \right]
$$

$$
+ \frac{1}{4\pi} \int d^3 \vec{r} \left[ \vec{E} \cdot \left( -\frac{1}{c} \frac{\partial}{\partial t} \vec{A} - \vec{\nabla} \Phi \right) - \vec{B} \cdot \vec{\nabla} \times \vec{A} + \frac{1}{2} (B^2 - E^2) \right]. \quad (9.5)
$$

Here, in the first line, the behavior of the charged particles (electrons) under the influence of the electromagnetic field (E.M.F.) is described, while the second line contains the pure field contributions.

The independent variables of particle and field variables are listed in two groups:

$$
\vec{r}_k(t), \; \vec{v}_k(t), \; \vec{p}_k(t)
$$

$$
\Phi(\vec{r}, t), \; \vec{A}(\vec{r}, t), \; \vec{E}(\vec{r}, t), \; \vec{B}(\vec{r}, t).
$$

Now we need to study the response of the Lagrangian (9.5) with regard to each of these variables.

We begin with the contribution of the particle variables:

$$\delta \vec{r}_k : \delta L = \frac{d}{dt}\left(\delta \vec{r}_k \cdot \vec{p}_k\right) + \delta \vec{r}_k \cdot \left[-\frac{d\vec{p}_k}{dt} - \vec{\nabla}_k e_k (\Phi(\vec{r}_k, t) - \frac{\vec{v}_k}{c} \cdot \vec{A}(\vec{r}_k, t))\right]$$

$$\delta \vec{v}_k : \delta L = \delta \vec{v}_k \cdot \left[-\vec{p}_k + m_k \vec{v}_k + \frac{e_k}{c}\vec{A}(\vec{r}_k, t)\right]$$

$$\delta \vec{p}_k : \delta L = \delta \vec{p}_k \cdot \left(\frac{d\vec{r}_k}{dt} - \vec{v}_k\right).$$

As formulated by the stationary condition of the action principle, we obtain the following equations of motion:

$$\frac{d\vec{p}_k}{dt} = -e_k \vec{\nabla}_k \left(\Phi(\vec{r}_k, t) - \frac{\vec{v}_k}{c} \cdot \vec{A}(\vec{r}_k, t)\right),$$

$$m\vec{v}_k = \vec{p}_k - \frac{e_k}{c}\vec{A}(\vec{r}_k, t),$$

$$\vec{v}_k = \frac{d\vec{r}_k}{dt} \tag{9.6}$$

These equations are very similar to the ones in the introduction of this chapter.

The situation for the field-dependent parts is not quite so simple. Note that in the first part of (9.5) there are also field contributions, but with respect to the individual points $\vec{r}_k$. As can be seen from (9.5), we have to sum over the particle contribution while the field contribution has to be integrated over. So we first have to transform the field-dependent sum into an integration. This is easily done with the aid of the $\delta$ function:

$$e_k \Phi(\vec{r}_k, t) = \int d^3\vec{r}\, \delta(\vec{r} - \vec{r}_k) e_k \Phi(\vec{r}, t),$$

so that we can write

$$\sum_k e_k \Phi(\vec{r}_k, t) = \int d^3\vec{r} \sum_k e_k \delta(\vec{r} - \vec{r}_k(t)) \Phi(\vec{r}_k, t),$$

$$= \int d^3\vec{r}\, \varrho(\vec{r}, t)\Phi(\vec{r}, t)$$

$$\varrho(\vec{r}, t) = \sum_k e_k \delta(\vec{r} - \vec{r}_k(t))$$

Likewise, $\vec{j}(\vec{r}, t) = \sum_k e_k \vec{v}_k(t)\delta(\vec{r} - \vec{r}_k(t))$.

The term that describes the coupling between particles and fields is thus given by

$$\sum_k \left[ -e_k \Phi(\vec{r}_k, t) + \frac{e_k}{c} \vec{v}_k \cdot \vec{A}(\vec{r}_k, t) \right] = \int d^3\vec{r} \left[ -\rho(\vec{r}, t)\Phi(\vec{r}, t) + \frac{1}{c} \vec{j}(\vec{r}, t) \cdot \vec{A}(\vec{r}, t) \right]$$

$$= \int d^3\vec{r} \left[ -\Phi + \frac{1}{c}\vec{j} \cdot \vec{A} \right] \tag{9.7}$$

The volume of the volume integral should be chosen sufficiently large so as to. contain all the charge and current densities.

Finally, we can use the principle of stationary action to obtain with the variation of the various fields the following equations:

$$\delta\Phi : \delta L = \frac{1}{4\pi} \int d^3\vec{r}\, \delta\Phi \left[ \vec{\nabla} \cdot \vec{E} - 4\pi\rho \right],$$

$$\vec{\nabla} \cdot \vec{E} = 4\pi\rho \tag{9.8}$$

$$\delta\vec{A} : \delta L = -\frac{1}{4\pi c}\frac{d}{dt} \int d^3\vec{r}\, \vec{E} \cdot \delta\vec{A} + \frac{1}{4\pi} \int d^3\vec{r} \left[ \frac{1}{c}\frac{\partial}{\partial t}\vec{E} + \frac{4\pi}{c}\vec{j} - \vec{\nabla} \times \vec{B} \right] \cdot \delta\vec{A}$$

$$\vec{\nabla} \times \vec{B} = \frac{1}{c}\frac{\partial}{\partial t}\vec{E} + \frac{4\pi}{c}\vec{j} \tag{9.9}$$

$$\delta\vec{E} : \delta L = \frac{1}{4\pi} \int d^3\vec{r}\, \delta\vec{E} \cdot \left( -\frac{1}{c}\frac{\partial}{\partial t}\vec{A} - \vec{\nabla}\Phi - \vec{E} \right)$$

$$\vec{E} = -\vec{\nabla}\Phi - \frac{1}{c}\frac{\partial\vec{A}}{\partial t} \tag{9.10}$$

$$\delta\vec{B} : \delta L = \frac{1}{4\pi} \int d^3\vec{r}\, \delta\vec{B} \cdot (-\vec{\nabla} \times \vec{A} + \vec{B})$$

$$\vec{B} = \vec{\nabla} \times \vec{A} \tag{9.11}$$

Contained in these equations are Maxwell's equations, two of which provide us with the construction of $\vec{E}$ and $\vec{B}$ with the aid of the potentials $\Phi$ and $\vec{A}$.

It is important to notice that $\vec{E}$ and $\vec{A}$ satisfy equations of motion, but $\Phi$ and $\vec{B}$ do not!

Next, we want to study the time variation in order to identify the Hamiltonian:

$$\delta t : \delta W = \int dt \left[ \frac{d}{dt}(-H\delta t) + \delta t\frac{dH}{dt} \right]$$

$$H = \sum_k \left[ \vec{p}_k \cdot \vec{v}_k - \frac{1}{2}m_k v_k^2 + e_k\Phi(\vec{r}_k, t) - \frac{e_k}{c}\vec{v}_k \cdot \vec{A}(\vec{r}_k, t) \right]$$

$$+ \frac{1}{4\pi} \int d^3\vec{r} \left[ \vec{E} \cdot \vec{\nabla}\Phi + \vec{B} \cdot \vec{\nabla} \times \vec{A}) + \frac{1}{2}(E^2 - B^2) \right]. \tag{9.12}$$

We will show that $H$ is a constant of motion, $\frac{dH}{dt} = 0$. Furthermore, the generators of the total time derivative can be easily identified:

$$\delta W_{12} = G_1 - G_2$$

where

$$G = \sum_k \vec{p}_k \cdot \delta \vec{r}_k - \frac{1}{4\pi c} \int d^3\vec{r}\, \vec{E} \cdot \delta \vec{A} - H\delta t. \qquad (9.13)$$

The Hamiltonian (9.12) has a twofold significance. First, it provides us with Hamilton's equation of motion. Second, it gives us the total energy of the system. This can be shown with the aid of the following equations (which do not contain time derivatives). They are split up in particle and field energy:

$$\vec{p}_k = m_k \vec{v}_k + \frac{e_k}{c} \vec{A}(\vec{r}_k, t)$$

$$\vec{B} = \vec{\nabla} \times \vec{A}$$

$$\vec{\nabla} \cdot \vec{E} = 4\pi\rho.$$

Then it holds that

$$H = E = \sum_k \left( \frac{1}{2} m_k v_k^2 + e_k \Phi_k \right) \text{ particle energy} \qquad (9.14)$$

$$+ \int d^3\vec{r} \left[ \frac{E^2 + B^2}{8\pi} - \rho\Phi \right] \text{ field energy} \qquad (9.15)$$

Now we want to show that $E$ is conserved, i.e., $\frac{dE}{dt} = 0$.n
Earlier we found

$$\frac{d}{dt} \sum_k (\frac{1}{2} m_k v_k^2 + e_k \Phi_k) = \sum_k \frac{\partial}{\partial t} (e_k \Phi_k - \frac{e_k}{c} \vec{v}_k \cdot \vec{A}_k) \qquad (9.16)$$

So still need

$$\frac{d}{dt} \int d^3\vec{r} \left[ \frac{E^2 + B^2}{8\pi} - \rho\Phi \right]. \qquad (9.17)$$

The last expression can be rewritten by using the energy conservation theorem

$$\frac{\partial U}{\partial t} + \vec{\nabla} \cdot \vec{S} + \vec{j} \cdot \vec{E}, U = \frac{E^2 + B^2}{8\pi}. \qquad (9.18)$$

So we can continue to write for (9.15)

$$\int d^3\vec{r}\left[-\vec{j}\cdot\vec{E}-\frac{\partial}{\partial t}\Phi-\frac{\partial\Phi}{\partial t}\right]\quad\left(\frac{\partial}{\partial t}+\vec{\nabla}\cdot\vec{j}=0,\ \vec{E}=\frac{1}{c}\frac{\partial\vec{A}}{\partial t}-\vec{\nabla}\varphi\right)$$

$$=\int d^3\vec{r}\left[-\vec{j}\cdot\vec{E}+\left(\vec{\nabla}\cdot\vec{j}\right)\Phi-\frac{\partial\Phi}{\partial t}\right]((\vec{\nabla}\cdot\vec{j})\Phi=\vec{\nabla}\cdot\left(\Phi\vec{j}\right)-\vec{j}\cdot\vec{\nabla}\Phi)\text{ surface term}$$

$$=\int d^3\vec{r}\left[\vec{j}\cdot\frac{1}{c}\frac{\partial\vec{A}}{\partial t}+\vec{j}\cdot\vec{\nabla}\Phi-\vec{j}\cdot\vec{\nabla}\Phi-\frac{\partial\Phi}{\partial t}\right]$$

$$=-\int d^3\vec{r}\left[\rho\frac{\partial\Phi}{\partial t}-\frac{1}{c}\cdot\frac{\partial\vec{A}}{\partial t}\right]$$

$$=-\sum_k e_k\left[\frac{\partial}{\partial t}\Phi_k-\frac{1}{c}\vec{\nabla}_k\cdot\frac{\partial}{\partial t}\vec{A}_k\right]=-\frac{d}{dt}\sum_k\left(\frac{1}{2}m_k v_k^2+e_k\Phi_k\right).$$

This term cancels exactly the initial particle contribution (9.14). Since the $\Phi$-dependent terms in $E$ cancel each other, it holds also that

$$E=\sum_k\frac{1}{2}m_k v_k^2+\int d^3\vec{r}\,\frac{E^2+B^2}{8\pi}.$$

We already found as a consequence of the action principle the generator

$$G=\sum_k\vec{p}_k\cdot\delta\vec{r}_k-\frac{1}{4\pi c}\int d^3\vec{r}\,\vec{E}\cdot\delta\vec{A}-\delta t E.\qquad(9.19)$$

This generator is going to provide a connection between conservation theorems and invariances of the action. In the following we want to derive the conservation of linear and angular or momentum from the invariance of the action under translational and rotational change of the coordinate system.

When changing the origin of a coordinate system by a rigid displacement $\delta\vec{r}$, a given point located relative to the old coordinate system by the vector $\vec{r}$ will be described in the new coordinate system by $\vec{r}+\delta\vec{r}$.

Now, the response of the particle term in $\delta W$ is simply $G_{particle}=\delta\vec{r}\cdot\sum_k\vec{p}_k$.

The field contribution $\delta\vec{A}$ is a little more difficult. But it can be found if we realize that generally the value of a function F at an arbitrary point $P$ remains unchanged with regard to the coordinates in the two respective systems, $F$ and $\overline{F}$ exists so that

$$F(p)=F(\vec{r})=\overline{F}(\overline{\vec{r}})=\overline{F}(\vec{r}+\delta\vec{r})\qquad(9.20)$$

Therefore the change of the function $F$, which is defined by

$$\overline{F}(\vec{r})=F(\vec{r})+\delta F(\vec{r})$$
$$\text{is given by }\delta F(\vec{r})=F(\vec{r}-\delta\vec{r})-F(\vec{r})$$
$$=-\delta\vec{r}\cdot\vec{\nabla}F(\vec{r}).\qquad(9.21)$$

(This holds true for a rigid displacement $\delta \vec{r}$, not for rotation.)
With the relation (9.21), the field contribution of $G$ is given by

$$-\frac{1}{4\pi c} \int d^3\vec{r}\, \vec{E} \cdot \delta \vec{A} = \frac{1}{4\pi c} \int d^3\vec{r}\, E_i \left(\delta \vec{r} \cdot \vec{\nabla}\right) A_i$$

$$= -\frac{1}{c} \sum_k e_k \delta \vec{r} \cdot \vec{A}_k + \frac{1}{4\pi c} \int d^3\vec{r}\, \left(\vec{E} \times \vec{B}\right) \cdot \delta \vec{r},$$

where we used.

$$\left(\delta \vec{r} \cdot \vec{\nabla}\right) \vec{A} = \left(\vec{\nabla} \times \vec{A}\right) \times \delta \vec{r} + \vec{\nabla} \left(\delta \vec{r} \cdot \vec{A}\right), \ \vec{\nabla} \cdot \vec{E} = 4\pi\rho), \ \vec{\nabla} \times \vec{A} = \vec{B}.$$

Altogether we obtain for the generator $G$ the following expression:

$$G = \delta \vec{r} \cdot \vec{P}$$

with

$$\vec{P} = \sum_k (\vec{p}_k - \frac{e_k}{c} \vec{A}_k) + \frac{1}{4\pi c} \int d^3\vec{r} \left(\vec{E} \times \vec{B}\right)$$

$$= \sum_k m_k \vec{v}_k + \int d^3\vec{r}\, \vec{G}, \ \vec{G} = \frac{1}{4\pi c} \vec{E} \times \vec{B},$$

where the first equation is the particle contribution and $\vec{G}$ is the momentum density.
Since the action is translational invariant, we have

$$0 = \delta W_{12} = G_1 - G_2 = \left(\vec{P}_1 - \vec{P}_2\right) \cdot \delta \vec{r},$$

and consequently, $\vec{P}_1 = \vec{P}_2$, i.e., the total momentum $\vec{P}$, is conserved.
Similar arguments can be used when rotating the coordinate system:

$$\delta \vec{r} = \delta \vec{\omega} \times \vec{r}.$$

Such a rotation affects a vector function according to

$$\vec{A}(\vec{r} + \delta \vec{r}) = \vec{A}(\vec{r}) + \delta \vec{\omega} \times \vec{A}(\vec{r}),$$

so that the new vector function with reference to the old coordinates is given by

$$\overline{\vec{A}}\left(\overline{\vec{r}}\right) = \vec{A}\left(\vec{r}\right) - \left(\delta\vec{r}\cdot\vec{\nabla}\right)\vec{A}\left(\vec{r}\right) + \delta\vec{\omega}\times\vec{A}\left(\vec{r}\right),$$

which implies the change in the vector potential

$$\delta\vec{A} = -\left(\delta\vec{r}\cdot\vec{\nabla}\right)\vec{A}\left(\vec{r}\right) + \delta\vec{\omega}\times\vec{A}.$$

Inserted in the generator, we obtain the following form:

$$G = \delta\vec{\omega}\cdot\vec{J},$$

where the total angular momentum $\vec{J}$ is given by

$$\vec{J} = \sum_k \vec{r}_k \times m_k \vec{v}_k + \int d^3\vec{r}\,\vec{r}\times\left(\frac{1}{4\pi c}\vec{E}\times\vec{B}\right), \qquad (9.22)$$

with $\frac{d\vec{J}}{dt} = 0$, i.e., $\vec{J}$ is conserved.

Electromagnetic fields possess a further invariance, namely, under gauge transformation:

$$\vec{A} \to \vec{A} - \vec{\nabla}\lambda$$
$$\Phi \to \Phi + \frac{1}{c}\frac{\partial}{\partial t}\lambda(\vec{r}, t). \qquad (9.23)$$

It is interesting to see how the Lagrangian (9.5) responds under a gauge transformation. Trivially, the pure field-dependent part remains unchanged. In the particle term we would like to have the velocity invariant under gauge transformation. Given our earlier finding,

$$m\vec{v} = \vec{p} - \frac{e}{c}\vec{A},$$

we can have $m\vec{v}$ invariant, if, besides $\vec{A} \to \vec{A} - \vec{\nabla}\lambda$, also $\vec{p}$ changes as $\vec{p} \to \vec{p} - \frac{e}{c}\vec{\nabla}\lambda$.

Then the Lagrangian changes as follows:

$$L \to \overline{L} \equiv L + \sum_k\left[-\frac{e_k}{c}\vec{\nabla}\lambda\cdot\left(\frac{d\vec{r}_k}{dt} - \vec{v}_k\right) - \frac{e_k}{c}\frac{\partial}{\partial t}\lambda - \frac{e_k}{c}\vec{v}_k\cdot\vec{\nabla}\lambda\right]$$
$$= L - \sum_k\left[\frac{e_k}{c}\left(\frac{\partial}{\partial t} + \frac{d\vec{r}_k}{dt}.\vec{\nabla}\right)\lambda\right] = L - \sum_k\frac{e_k}{c}\frac{d\lambda}{dt}$$

$$= L - \frac{dw}{dt} \tag{9.24}$$

with

$$w = \sum_k \frac{e_k}{c} \lambda(\vec{r}_k, t) \tag{9.25}$$

Since the addition of a total time derivative (9.24) does not change the equations of motion, the whole system stays invariant. But the end point behavior is changed:

$$\overline{W}_{12} = W_{12} - (w_1 - w_2)$$

so that we obtain a new generator,

$$\overline{G} = G - w. \tag{9.26}$$

To understand the consequence of gauge invariance, let us consider a special case for the gauge function $\lambda(\vec{r}, t)$, namely,

$$\lambda(\vec{r}, t) = const. = \delta\lambda.$$

Since the potential $\vec{A}$ and $\Phi$ as well as $\vec{p}_k$ remain unchanged, we have $\overline{L} = L$. Consequently, we obtain

$$\frac{\delta\lambda}{c} \frac{d}{dt} \left( \sum_k e_k \right) = 0 \text{ or } \sum_k e_k = Q, \tag{9.27}$$

which means gauge invariance implies charge conservation. Also, since Maxwell's equations are invariant under Lorentz transformation, $Q$ is not just a scalar; it is a Lorentz scalar.

# References

1  Dittrich, W., Reuter, M.: Classical and Quantum Dynamics, 6th edn. Springer, Berlin (2020)
2. Schwinger, J., Milton, K.A., DeRaad, L.L., Tsai, W.-Y.: Classical Electrodynamics. Westview Press, Cambridge, MA, USA (1998)

# Chapter 10
# The Two Giants in Gravity: Einstein and Hilbert

This chapter is devoted to one of the most spectacular achievements of the 20th century. On November 25, 1915, Einstein found the following famous equation (See page 522 in chapter 41 of [1] and Weinberg's book [2]):

$$R_{\mu\nu} = \kappa \left( T_{\mu\nu} - \frac{1}{2} g_{\mu\nu} T \right), \kappa = 8\pi G. \tag{10.1}$$

Practically simultaneously, Hilbert arrived at the same result using the action principle. The question of priority does not come up here, since both authors used entirely different methods, so that one could not be completely certain that they would reach the same result. But it is pretty fair to say that Einstein is the true creator of the physical theory of general relativity, while concurrently, Einstein and Hilbert share the fame for the discovery of the fundamental Eq. (10.1).

In the following, we will pursue both paths, beginning with Einstein's search for a tensor that can reproduce the classical limit of (10.1) formulated by the Newton–Poisson equation $\vec{\nabla}^2 \Phi(\vec{r}) = 4\pi G\rho$, which is one of the tests of the validity of (10.1).

To prove Eq. (10.1), we start with the following definition that fixes the Newtonian limit:

$$R_{00} = \vec{\nabla}^2 \Phi = 4\pi G\varrho = 4\pi G T_{00}. \tag{10.2}$$

Then we introduce the important tensor $G_{\mu\nu}$ :

$$G_{\mu\nu} = R_{\mu\nu} - \frac{1}{2} g_{\mu\nu} R \tag{10.3}$$

© The Author(s), under exclusive license to Springer Nature Switzerland AG 2021
W. Dittrich, *The Development of the Action Principle*,
SpringerBriefs in Physics,
https://doi.org/10.1007/978-3-030-69105-9_10

and compute the various components of (10.3):

$$G_{00} = R_{00} - \frac{1}{2}g_{00}\left(R_0^0 + R_j^j\right) = \kappa\varrho$$
$$G_i^i = R_i^i - \frac{1}{2}\left(R_0^0 + R_j^j\right) = 0 \quad (no\,summation)$$

Taking the sum over $i$ we have

$$G_i^i = R_i^i - \frac{3}{2}\left(R_0^0 + R_j^j\right)$$
$$= -\frac{3}{2}R_0^0 - \frac{1}{2}R_j^j = 0 : R_j^j = -3R_0^0.$$

Thus we can write for (10.3):

$$G_{00} = R_{00} - \frac{1}{2}g_{00}\left(R_0^0 - 3R_0^0\right)$$
$$= R_{00} + g_{00}R_0^0 = 2R_0^0 = \kappa\varrho$$

Here we substitute Eq. (10.2) so that

$$G_{00} = 2R_{00} = \kappa\varrho = 2\vec{\nabla}^2\Phi = 2\cdot 4\pi G\rho = 8\pi GT_{00} = \kappa T_{00}.$$

This result can be generalized to

$$G_{\mu\nu} = 8\pi GT_{\mu\nu} = \kappa T_{\mu\nu}, \tag{10.4}$$

which is equivalent to Einstein's Eq. (10.1).

This can be seen as follows:

$$\left(G_{\mu\nu} =\right)R_{\mu\nu} - \frac{1}{2}g_{\mu\nu}R = \kappa T_{\mu\nu} \tag{10.5}$$

$$R_\mu^\nu - \frac{1}{2}\delta_\mu^\nu R = \kappa T_\mu^\nu$$
$$R - 2R = \kappa T : -R = \kappa T. \tag{10.6}$$

When we insert $R = -\kappa T$ in (10.5), we find

$$R_{\mu\nu} - \frac{1}{2}g_{\mu\nu}R = R_{\mu\nu} + \frac{1}{2}g_{\mu\nu}\kappa T = \kappa T_{\mu\nu}$$
$$R_{\mu\nu} = \kappa\left(T_{\mu\nu} - \frac{1}{2}g_{\mu\nu}T\right), \kappa = 8\pi G,$$

which brings us back to (10.1).

Let us keep in mind that both expressions,

$$G_{\mu\nu} = \kappa T_{\mu\nu}$$

and

$$R_{\mu\nu} = \kappa \left( T_{\mu\nu} - \frac{1}{2} g_{\mu\nu} T \right), \tag{10.7}$$

are equivalent versions of the Einstein–Hilbert field equation for the metric field $g_{\mu\nu}$, or, better, for the 10 potentials $g_{\mu\nu}$.

It is interesting to observe that Einstein's equations are unique. To demonstrate this, we begin with the most general ansatz,

$$\mathcal{D}_{\mu\nu}[g] = 8\pi G T_{\mu\nu}. \tag{10.8}$$

$\mathcal{D}_{\mu\nu}[g]$ is a tensor functional which is restricted by the following conditions:

(a)   $\mathcal{D}_{\mu\nu}[g]$ depends only upon $g_{\mu\nu}, g_{\mu\nu,\alpha}, g_{\mu\nu,\alpha\beta}$. This is required to respect the Newtonian limit, which is a second-order partial differential equation;
(b)   $\mathcal{D}^{\mu\nu}{}_{;\nu} = 0$, because $T^{\mu\nu}{}_{;\nu} = 0$;
(c)   $\mathcal{D}^{\mu\nu}$ is linear in second derivatives.

If we limit ourselves to 4 dimensions, then the only tensors which satisfy (a) and (b) are $G_{\mu\nu}$ and $g_{\mu\nu}$. Therefore, the most general form of $\mathcal{D}_{\mu\nu}[g]$ is given by

$$\mathcal{D}_{\mu\nu}[g] = aG_{\mu\nu} + bg_{\mu\nu},$$

and since $G^{\mu\nu}{}_{;\nu} = 0 = g^{\mu\nu}{}_{;\nu}$, i.e., their covariant derivatives are zero, we obtain indeed $\mathcal{D}^{\mu\nu}{}_{;\nu} = 0$. So we have decomposed $\mathcal{D}_{\mu\nu}$ into two separate functions, $g_{\mu\nu}$ and $G_{\mu\nu}$, which remain unchanged under general coordinate transformations.

Finally, the Einstein–Hilbert field equations take the two equivalent forms

$$G_{\mu\nu} - \Lambda g_{\mu\nu} = \kappa T_{\mu\nu}, \tag{10.9}$$

$$R_{\mu\nu} - \Lambda g_{\mu\nu} = \kappa \left( T_{\mu\nu} - \frac{1}{2} g_{\mu\nu} T \right). \tag{10.10}$$

Here, $\Lambda$ is the cosmological constant and $\kappa = 8\pi G$.

We now turn to the derivation of the Hilbert–Einstein field equations via the action principle, i.e., we set up the Lagrangian formulation for general relativity. Then the variation $\delta S_{gr} = 0$ should give us the E.–H. equations. For gravity we define the action functional

$$S_{gr} = \int d^4x \sqrt{-g}\, \mathcal{L}_{gr}. \tag{10.11}$$

As it turns out, to produce the correct E.–H. equations, we simply have to choose $\mathcal{L}_{gr} = R$. Hence we have to work out the variation of the following action:

$$S_{gr} = \int d^4x \sqrt{-g}\, R. \tag{10.12}$$

The principle of stationary action tells us that we have to calculate the following variations ($R = g^{\mu\nu} R_{\mu\nu}$) :

$$0 = \delta S_{gr} = \delta \int d^4x \sqrt{-g}\, R = \int d^4x (\delta \sqrt{-g}) \cdot g^{\mu\nu} R_{\mu\nu} + \sqrt{-g} \cdot \delta g^{\mu\nu} R_{\mu\nu} + \sqrt{-g} g^{\mu\nu} \delta R_{\mu\nu}). \tag{10.13}$$

As can be seen, we are in need of the following useful relations (which are not proven here):

(1)    (1)$\delta g^{\mu\nu} = -g^{\mu\alpha} g^{\nu\beta} \delta g_{\alpha\beta}$
(2)    $\delta \sqrt{-g} = \frac{1}{2} \sqrt{-g}\, g^{\alpha\beta} \delta g_{\alpha\beta}$
(3)    $\delta R_{\mu\nu} \cdot g^{\mu\nu} = \omega^{\alpha}{}_{;\alpha}$ for some appropriate $\omega^{\alpha}$, "divergence"
(4)    $\omega^{\alpha}{}_{;\alpha} = \frac{1}{\sqrt{-g}} \frac{\partial}{\partial x^{\mu}} (\sqrt{-g}\, \omega^{\mu})$, for any vector field
(5)    $\int d^4x \sqrt{-g}\, \omega^{\alpha}{}_{;\alpha} = 0$, if $\omega^{\alpha} = 0$ at infinity; Gauss' theorem, i.e.,
        $\int d^4x \sqrt{-g}\, \omega^{\alpha}{}_{;\alpha} = \int d^4x \frac{\partial}{\partial x^{\mu}} (\sqrt{-g}\, \omega^{\mu}) = \oint d^3 S_{\mu} \sqrt{-g}\, \omega^{\mu} = 0$.

Surface at infinity
Given these formulae, we can continue with (10.13) and so obtain

$$0 = \delta S_{gr} = \int d^4x \left[ \frac{1}{2} \sqrt{-g}\, g^{\alpha\beta} g^{\mu\nu} R_{\mu\nu} \delta g_{\alpha\beta} - \sqrt{-g} g^{\mu\alpha} g^{\nu\beta} R_{\mu\nu} \delta g_{\alpha\beta} + (\sqrt{-g}) \omega^{\alpha}{}_{;\alpha} = 0) \right]$$

$$= \int d^4x \left( \frac{1}{2} g^{\alpha\beta} g^{\mu\nu} R_{\mu\nu} - g^{\mu\alpha} g^{\nu\beta} R_{\mu\nu} \right) \delta g_{\alpha\beta}$$

$$= -\int d^4x (R^{\alpha\beta} - \frac{1}{2} g^{\alpha\beta} R) \delta g_{\alpha\beta}.$$

So for pure gravity we obtain, indeed

$$R^{\alpha\beta} - \frac{1}{2} g^{\alpha\beta} R = 0 \tag{10.14}$$

If we include matter, we have to use the action principle for

$$S = S_{gr} + S_{matter} = k \int d^4x \sqrt{-g}\, R + \int d^4x \sqrt{-g}\, \mathcal{L}_{matter}.$$

Here we need

$$\delta g_{\alpha\beta} S = \delta g_{\alpha\beta} S_{gr} + \delta_{\alpha\beta} S_{matter} = 0$$

The result is

$$S = -\frac{1}{16\pi G} \int d^4x \sqrt{-g} R + \int d^4x \sqrt{-g} \mathcal{L}_{matter}.$$

Note the definition of $T^{\mu\nu}$:

$$\delta S_{matter} = -\frac{1}{2} \int d^4x \sqrt{-g} T^{\mu\nu} \delta g_{\mu\nu}$$

and

$$G^{\mu\nu} = -\frac{1}{2k} T^{\mu\nu} = 8\pi G T^{\mu\nu} : k = -\frac{1}{16\pi G}.$$

We finally want to consider an astronomical problem that had been unresolved since Newton's times. We now know that the force experienced by a planet in the solar system is not exerted only by the sun, but also by all the other planets. Hence the gravitational force between the sun and the planet alone, which varies with $\frac{1}{r^2}$, becomes modified.

Detailed calculations of the influence of the other planets on the orbital motion of Mercury predicted that the rate of advance of the perihelion should be approximately 531″ per century compared to the measured effects of 574″ per arc per century. So there remained a difference of 43″ between observation and calculation.

Meanwhile, due to Einstein, general relativity almost exactly accounts for the difference of 43″. This result was one of the first major triumphs of the theory of general relativity. Einstein himself described this achievement as the most intensive experience he had ever met in his scientific life, maybe in his entire life: "Ich war einige Tage fassungslos vor Erregung." ["For a few days my excitement was overwhelming".]

One should not forget that in 1916, when Einstein discovered the anomaly of Mercury's path, nothing was known about the red shift, and light deflection was detected only in 1919.

Now we want to start with Newton's law, which tells us how the planet Mercury moves under the influence of the gravitational inverse-square law:

$$F(r) = -G\frac{mM}{r^2}. \tag{10.15}$$

When we combine the equations that express the conservation of energy and angular momentum, we obtain the equation of the orbit $\theta(r)$ :

$$\frac{d^2}{d\theta^2}\left(\frac{1}{r}\right) + \frac{1}{r} = -\frac{mr^2}{l^2}F(r) \tag{10.16}$$

or, by taking a simple change of variable, $u = \frac{1}{r}$,

$$\frac{d^2u}{d\theta^2} + u = -\frac{m}{l^2}\frac{1}{u^2}F\left(\frac{1}{u}\right) \tag{10.17}$$

$$F\left(\frac{1}{u}\right) = F(r) = -G\frac{mM}{r^2} = -GmMu^2$$

The modification of the gravitational-force law requires by the general theory of relativity the addition of a new force component that varies as $\frac{1}{r^4} = u^4$. Therefore, Eq. (10.17) becomes modified according to

$$-\frac{m}{l^2}\frac{1}{u^2}F\left(\frac{1}{u}\right) = \frac{Gm^2M}{l^2} + \frac{3GM}{c^2}u^2, \tag{10.18}$$

where the contributions on the right-hand side are due to Newton and Einstein.

So the planet experiences the total force,

$$F\left(\frac{1}{u}\right) = -GmMu^2 - \frac{3GMl^2}{mc^2}u^4,$$

or

$$F(r) = -GmM\frac{1}{r^2} - \frac{3GMl^2}{mc^2}\frac{1}{r^4},$$

$$= -\frac{\partial V}{\partial r} \tag{10.19}$$

so that we obtain

$$V(r) = -\frac{GmM}{r} - \frac{GMl^2}{mc^2}\frac{1}{r^3}. \tag{10.20}$$

After having identified the new law of force and the associated potential, we can write down a Hamiltonian for our interacting system of the sun and Mercury. This brings us back to Hamiltonian mechanics with

$$H(r, p) = \frac{p^2}{2m} - \frac{GmM}{r} - \frac{GMl^2}{mc^2}\frac{1}{r^3} \tag{10.21}$$

or

$$H = H_{grav.} + H'$$

with

$$H_{grav.} = \frac{p^2}{2m} - \frac{k}{r}, k = GmM$$

and

$$H' = -\frac{\lambda}{r^3}, \lambda = \frac{GMl^2}{mc^2} \tag{10.22}$$

Without the $\frac{1}{r^3}$ term, we find for the equations of motion:

$$\frac{d\vec{r}}{dt} = \frac{\partial H_{grav.}}{\partial \vec{p}} = \frac{\vec{p}}{m}, \frac{d\vec{p}}{dt} = -\frac{\partial H_{grav.}}{\partial r} = -\frac{k}{r^3}\vec{r},$$

For the angular momentum we have $\vec{L} = \vec{r} \times \vec{p}$, which is conserved:

$$\frac{d\vec{L}}{dt} = 0.$$

Furthermore, $\vec{p} \cdot \vec{L} = 0$.

A few more lines of calculations (all steps are known from mechanics) are needed to arrive at.

$$\frac{d}{dt}\left(\frac{1}{km}\vec{L} \times \vec{p}\right) = -\frac{d\vec{r}}{dt r}. \tag{10.23}$$

At this stage we bring in the famous Laplace–Runge–Lenz–Pauli vector $\vec{A}$ ([3, 4]):

$$\vec{A} = \frac{1}{mk}\vec{L} \times \vec{p} + \frac{\vec{r}}{r}, \frac{d\vec{A}}{dt} = 0. \tag{10.24}$$

As is well known, the existence of this additional conserved vector together with H and $\vec{L}$ allows us to solve the entire Kepler problem and, in quantum mechanics, is responsible for the $n^2$ degeneracy in the hydrogen atom with $\frac{1}{r}$ potential.

Now we proceed via Poisson brackets and use the Hamiltonian (10.22):

$$H = H_{grav.} - \frac{\lambda}{r^3}.$$

As a first result we note that $\left\{\vec{A}, H_{grav.}\right\} = 0$, since $\frac{d\vec{A}}{dt} = 0$ for a pure $\frac{1}{r}$ potential.

The more interesting bracket is, however:

$$
\begin{aligned}
\frac{d\vec{A}}{dt} = \left\{\vec{A}, H\right\} &= \left\{\vec{A}, \left(H_{grav.} - \frac{\lambda}{r^3}\right)\right\} \\
&= -\frac{\lambda}{mk}\left\{\vec{L} \times \vec{p}, \frac{1}{r^3}\right\} \\
&= -\frac{3\lambda}{mk}\frac{1}{r^5}\left(\vec{r} \times \vec{L}\right)
\end{aligned}
\tag{10.25}
$$

Furthermore,

$$
\vec{A} \times \vec{L} = \frac{1}{mk}L^2\vec{p} + \frac{\vec{r} \times \vec{L}}{r}.
\tag{10.26}
$$

Now, since most of the planets deviate only slightly from a circular trajectory, we want to average over the circular rather than the elliptic path:

$$
< r >= r(= a)
$$

Then it is obvious that $< \vec{p} >= 0$.

As a consequence, we can write using (10.26),

$$
< \vec{A} \times \vec{L} >= - < \vec{L} \times \vec{A} >=< \frac{\vec{r} \times \vec{L}}{r} > .
$$

From Eq. (10.25) we obtain Fig. 10.1

$$
\left\langle\frac{d\vec{A}}{dt}\right\rangle = \frac{3\lambda}{mk}\left\langle\frac{\vec{r} \times \vec{L}}{r^5}\right\rangle = -\frac{3\lambda}{mk}\frac{1}{r^4}\left\langle\vec{L} \times \vec{A}\right\rangle.
$$

**Fig. 10.1** $< \vec{p} >$ average over circular path

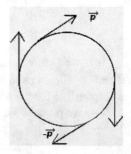

$$\frac{d\vec{A}}{dt} = \vec{\omega} \times \vec{A}: \quad \omega_p = \frac{3\lambda l}{mk}\frac{1}{r^4} = \frac{3GMl^2 l}{mc^2 mGmM}\frac{1}{r^4} = 3\frac{(mGmMr)^{\frac{3}{2}}}{mm^2c^2}\frac{1}{r^4}$$

$$= 3\frac{(GM)^{\frac{3}{2}}}{c^2}\frac{r^{\frac{3}{2}}}{r^4} = 3\frac{(GM)^{\frac{3}{2}}}{c^2}\frac{1}{r^{\frac{5}{2}}}.$$

Finally we obtain

$$\Delta\Theta = T\omega_p = \frac{2\pi r^{\frac{3}{2}}}{G^{\frac{1}{2}}M^{\frac{1}{2}}}3\frac{(GM)^{\frac{3}{2}}}{c^2}\frac{1}{r^{\frac{5}{2}}}$$

$$= \frac{6\pi GM}{rc^2}$$

$$\varepsilon \neq 0, r \to a(1-\varepsilon^2) \to \frac{6\pi GM}{ac^2(1-\varepsilon^2)}. \qquad (10.27)$$

For the planet Mercury, we have

$$\varepsilon = 0.206$$

$$a = 0.387 A.U. = 0.387 \cdot 1.495 \cdot 10^{11}\text{m}$$

Mass of the sun: $M = 1.99 \cdot 10^{30} kg$
Grav. constant: $G = 6.670 \cdot 10^{-11}\frac{m^3}{kgsec^2}$
Light velocity: $c = 2.998 \cdot 10^8 \frac{m}{sec}$
After inserting these numbers into Eq. (10.27), we obtain

$$\Delta\Theta = \frac{6\pi GM}{ac^2(1-\varepsilon^2)} = 5.066 \cdot 10^{-7} \text{ rad perrevolution.}$$

Mercury time for one revolution: 0.241 years.
The total precession of the perihelion of Mercury in one century is thus:

$$\Delta\Theta \cdot \frac{100}{0.241} = 2.1025 \cdot 10^{-4} rad = 43.4'' \text{ percentury.} \qquad (10.28)$$

With regard to Eq. (10.24), let us go back and reminisce about the usefulness of the polar vector $\vec{A}$ and refresh our memory of its role in classical mechanics and astronomy, i.e., in the Kepler (Coulomb) problem. To this end, we begin with the behavior of a point particle in a $\frac{1}{r}$ – potential:

$$V(r) - \frac{k}{r}, k = \text{const.} \qquad (10.29)$$

Next, we want to prove that except for the angular momentum $\vec{L}$, there is another conserved quantity, namely the Lenz vector $\vec{A}$.

Here are some of the necessary elementary mathematical steps to prove our claim:

$$\frac{d}{dt}\left(\vec{L} \times \vec{p}\right) = \vec{L} \times \frac{d\vec{p}}{dt} \quad \text{with} \frac{d\vec{L}}{dt} = 0$$

$$= \vec{L} \times \left(-\vec{\nabla}V\right) = \vec{L} \times \left(-\frac{k}{r^3}\vec{r}\right) = -k\left[(\vec{r} \times \vec{p}) \times \frac{\vec{r}}{r^3}\right]$$

$$= km\left[\frac{\vec{p}/m \cdot \vec{r}}{r^3}\vec{r} - \frac{\vec{p}}{m}\frac{1}{r}\right].$$

This equation can be written as

$$\frac{d}{dt}\left(\frac{1}{mk}\vec{L} \times \vec{p}\right) = \left(\frac{d}{dt}\vec{r}\right) \cdot \frac{\vec{r}}{r^3}\vec{r} - \left(\frac{d}{dt}\vec{r}\right)\frac{1}{r} = -\frac{d}{dt}\frac{\vec{r}}{r},$$

which brings us to the desired expression we were looking for, namely:

$$\vec{A} = \frac{1}{mk}\vec{L} \times \vec{p} + \frac{\vec{r}}{r} \text{ with } \frac{d}{dt}\vec{A} = 0.$$

This tells us that $\vec{L}$ as well as $\vec{A}$ are conserved quantities.

In the spirit of our action-principle approach, let us repeat the derivation of Eq. (10.24) and write for the action $W$

$$W_{1,2} = \int_{t_2}^{t_1} dt\left[p_i\frac{dx_i}{dt} - \frac{p_i^2}{2m} + \frac{k}{r}\right]. \tag{10.30}$$

The special form of the variations of $\delta x_i$ and $\delta p_i$ is now given by the rigid displacements ($\delta\varepsilon_i = \text{const.}, \delta t(t_{1,2}) = 0$) :

$$\delta x_j = \delta\varepsilon_i\left[\frac{1}{mk}\delta_{ij}x_k p_k - x_i p_j - \varepsilon_{ijk}L_k\right], L_k = \varepsilon_{klm}x_l p_m \tag{10.31}$$

$$\delta p_j = \delta\varepsilon_i\left[\frac{x_i x_j}{r^3} - \delta_{ij}\frac{1}{r} - \frac{1}{mk}(p_i p_j - \delta_{ij}p^2)\right]. \tag{10.32}$$

Here, in contrast to our former examples, $\delta p_i$ is no longer arbitrary. The calculation of $\delta W$ with the help of (10.31) and (10.32) is performed in the usual way and yields, after a few steps (here is an exercise):

$$\delta W = 2\delta\varepsilon_i\int_{t_2}^{t_1} dt\left(-\frac{d}{dt}\right)\left(\frac{x_i}{r}\right).$$

So the action principle then reads:

$$\delta W = -2\delta\varepsilon_i \int_{t_2}^{t_1} dt\, \frac{d}{dt}\left(\frac{x_i}{r}\right) = \int_{t_2}^{t_1} dt\, \frac{d}{dt}\, p_i \delta x_i.$$

For the integrand on the right-hand side we get

$$p_i \delta x_i = \frac{2}{mk}\delta\varepsilon_i\left[p_k x_k p_i - p^2 x_i\right] = \frac{2}{mk}\delta\varepsilon_i(\vec{L} \times \vec{p})_i.$$

Our final result is, therefore:

$$2\delta\vec{\varepsilon} \cdot \int_{t_2}^{t_1} dt\, \frac{d}{dt}\left[\left(-\frac{d}{dt}\left(\frac{\vec{r}}{r}\right)\right) - \frac{d}{dt}\frac{1}{mk}(\vec{L} \times \vec{p})\right] = 0.$$

So we have again proved that the Runge–Lenz vector $\vec{A}$ is a conserved quantity in the Coulomb (Kepler) problem:

$$\vec{A} := \frac{1}{mk}\vec{L} \times \vec{p} + \frac{\vec{r}}{r} \quad : \quad \frac{d\vec{A}}{dt} = 0.$$

We are now prepared to introduce the Poisson brackets, which can be used to show that the action $W_{1,2}$ under the infinitesimal changes, $\delta r_i$ and $\delta p_j$, remains invariant and thus produces as conserved quantity the Lenz vector. So under discussion is the Hamiltonian action principle, where by definition we have $W_{1,2} = \int_2^1 dt\left[\vec{p} \cdot \frac{d\vec{r}}{dt} - H(\vec{r}, \vec{p})\right]$, with the Hamiltonian given by $H(\vec{r}, \vec{p}) = \frac{\vec{p}^2}{2m} - \frac{k}{r}$.

From here follows the conservation of the Lenz vector by looking at the infinitesimal changes:

$$\delta r_j = \varepsilon_i\left[r_j, A_i\right]_{P.B.}, \delta p_j = \varepsilon_i\left[p_j, A_i\right] = m\delta\dot{r}_j.$$

The specific form of the variation for $\delta\vec{r}$ and $\delta\vec{p}$ follows from the explicit representation of the following Poisson brackets:

$$\left[A_i, r_j\right] = \frac{1}{mk}\left(\varepsilon_{ijk}L_k + r_i p_j - \delta_{ij}\vec{r} \cdot \vec{p}\right)$$

$$\left[A_i, p_j\right] = \frac{1}{mk}\left(p_i p_j - \delta_{ij}p^2\right) + \delta_{ij}\frac{1}{r} - \frac{r_i r_j}{r^2}$$

Using

$$\vec{p} \cdot \frac{d\vec{r}}{dt} - H(\vec{r}, \vec{p}) = m\frac{d\vec{r}}{dt} \cdot \frac{d\vec{r}}{dt} - \frac{m}{2}\left(\frac{d\vec{r}}{dt}\right)^2 + \frac{k}{r} = \frac{m}{2}\left(\frac{d\vec{r}}{dt}\right)^2 + \frac{k}{r} = T - V = L$$

or $L = \frac{m}{2}\dot{r}_j\dot{r}_j + \frac{k}{r}$, which is needed in $\delta W_{12} = \int_{t_2}^{t_1}dt\delta L$.
This means that we are required to find $\delta L$.
In order to save space, we only give the final result for $\delta L$:

$$\delta L = m\dot{r}_j\delta\dot{r}_j - k\frac{r_j}{r^3}\delta r_j = 2\varepsilon_i\left(-\frac{d}{dt}\right)\left(\frac{r_i}{r}\right).$$

Therefore, the action principle reads:

$$\delta W_{12} = \int_{t_2}^{t_1}dt\delta L = G_1 - G_2 = \int_{t_2}^{t_1}dt\frac{d}{dt}(\vec{p}\cdot\delta\vec{r}).$$

For the integrand in the surface term, we obtain

$$(\vec{p}\cdot\delta\vec{r}) = \frac{2}{mk}\varepsilon_i(\vec{L}\times\vec{p})_i.$$

Our final result for $\delta W_{12} = 0$ reproduces, as expected,

$$2\vec{\varepsilon}\cdot\int_{t_2}^{t_1}dt\left[\left(-\frac{d}{dt}\left(\frac{\vec{r}}{r}\right) - \frac{d}{dt}\frac{1}{mk}(\vec{L}\times\vec{p})\right)\right] = 0,$$

which means that the Runge–Lenz vector is indeed a conserved quantity:
$\frac{d\vec{A}}{dt} = 0$ with $\vec{A} = \frac{1}{mk}\left(\vec{L}\times\vec{p}\right) + \frac{\vec{r}}{r}$.

# References

1. Dittrich, W., Reuter, M.: Classical and Quantum Dynamics, 6th edn. Springer, Berlin (2020)
2. Weinberg, S.: Gravitation and Cosmology. Wiley, Berlin (1972)
3. Lenz, W.: Zs.f. Phys. **24**, 197 (1924)
4. Pauli, W.: Zs.f. Phys. **36**, 336 (1926)

# Chapter 11
# The Quantum Action Principle

Let us start with a note on the Quantum **Dynamical Principle** that Schwinger published. Here is a shortened excerpt of what he had to say (in 1953 [1]):

The author has developed a formulation of quantum mechanics which enables one to deduce all properties of a quantum mechanical system from a single dynamical principle. It is the subject of quantum dynamics to construct transformation functions of the type $\langle a'_1 t_1 | a''_2 t_2 \rangle$ where $a_1$ and $a_2$ are complete sets of commuting Hermitian operators. The fundamental dynamical principle is a **differential** characterization of such transformation functions. Any infinitesimal change can be represented by

$$\delta < a'_1 t_1 \left| a''_2 t_2 \right> = \frac{i}{\hbar} < a'_1 t_1 \left| \delta W_{12} | a''_2 t_2 \right>$$

which is the definition of the infinitesimal **operator** $W_{12}$. The composition and reality properties of transformation functions imply that

$$\delta W_{13} = W_{12} + \delta W_{23}, \quad \delta W_{12}^{\dagger} = \delta W_{12}.$$

The basic postulate of the dynamical principle is that there exists a class of alterations for which the associated operators $\delta W_{12}$ are obtained by appropriate variations of a single operator, the action operator $W_{12}$,

$$\delta W_{12} = \delta [W_{12}].$$

According to the additivity property of the action operator it must possess the form

$$W_{12} = \int_{t_2}^{t_1} dt\, L(t)$$

where $L(t)$ is a Hermitian function of fundamental dynamical variables $x_a(t)$, in the infinitesimal neighbourhood of the time $t$ ...

So far, Schwinger's ideas.

© The Author(s), under exclusive license to Springer Nature Switzerland AG 2021
W. Dittrich, *The Development of the Action Principle*,
SpringerBriefs in Physics,
https://doi.org/10.1007/978-3-030-69105-9_11

In classical mechanics, Hamilton's equations of motion are deduced from an action principle. Now, according to Heisenberg, quantum mechanics is neither a wave- nor a particle mechanics. It is a theory of states and observables. So we may ask ourselves: Is there, without any reference to classical mechanics à la Lagrange and Hamilton, an action principle in quantum mechanics? Indeed, there is! Without going into all the details of deriving it, we obtain for the variation of the transformation amplitude $\langle q', t_1 | q'' t_2 \rangle$ the Quantum Action Principle:

$$\delta\langle q', t_1 | q'', t_2 \rangle = \frac{i}{\hbar}\langle q', t_1 | \delta W_{12} | q'', t_2 \rangle \qquad (11.1)$$

$$\text{with} \qquad W_{12} = \int_{t_1}^{t_2} dt\, L(t). \qquad (11.2)$$

The action operator $W$ has dimension of $\hbar$, and L is the Lagrange operator. Note that no classical ideas were used in obtaining this form.

The transition amplitude $\langle q', t_1 | q'', t_2 \rangle$ depends on the choice of final and initial state, specified by the vectors $\langle q', t_1 |$ and $| q'', t_2 \rangle$, and upon the form of the Hamiltonian operator that generates the time development. For a given Hamiltonian, the only freedom of change is of the initial and final states, for which we write:

$$\delta\langle q', t_1 | = \frac{i}{\hbar}\langle q', t_1 | G_1, \quad \delta | q'', t_2 \rangle = -\frac{i}{\hbar} G_2 | q'', t_2 \rangle. \qquad (11.3)$$

$$\text{So that we obtain} \quad \delta W_{12} = G_1 - G_2. \qquad (11.4)$$

This is the Principle of Stationary Action. It asserts that the infinitesimal variation of $W_{12}$—which according to (11.2) depends upon the variables at all values of $t$ between $t_1$ and $t_2$—involves only variations at the endpoints $t_1$ and $t_2$ and so is stationary with respect to variations at any intermediate time. One can then show that the equations of motion and the commutation relations follow from this simple, fundamental dynamical principle:

$$\frac{\hbar}{i}\frac{\partial}{\partial q'_\alpha}\langle q', t | = \langle q', t | p_\alpha(t), \quad \frac{1}{i\hbar}[q_\alpha, p_\alpha] = \frac{\partial q_\alpha}{\partial q_\beta} = \delta_{\alpha\beta}$$

From the reality property:

$$\delta\left(\langle a' | b'\rangle^*\right) = \delta\langle b' | a'\rangle = \frac{i}{\hbar}\langle b' | \delta W_{ba} | a'\rangle$$

$$\text{and} \quad \delta\left(\langle a' | b'\rangle^*\right) = \left(\delta\langle a' | b'\rangle\right)^* = -\frac{i}{\hbar}\langle a' | \delta W_{ab} | b'\rangle^* = -\frac{i}{\hbar}\langle b' | \delta W_{ab}^\dagger | a'\rangle, \qquad (11.5)$$

$$\text{we infer}: \qquad \delta W_{ab}^\dagger = -\delta W_{ba} = +\delta W_{ab}.$$

Therefore, $\delta W_{ab}$ is Hermitian.

This Hermiticity suggests that we use an infinitesimal unitary operator to generate variations of the state function: $= e^{\frac{i}{\hbar}G} \rightarrow \left(1 + \frac{i}{\hbar}G\right)$, where $G$ is infinitesimal and Hermitian.

$$\text{Then} \qquad \delta\langle a'| = \langle a'|(U_a - 1) = \frac{i}{\hbar}\langle a'|G_a$$

$$\text{and} \qquad \delta|b'\rangle = \left(U_b^{-1} - 1\right)|b'\rangle = -\frac{i}{\hbar}G_b|b'\rangle,$$

$$\text{so that} \quad \delta\langle a'b'\rangle = \left(\delta\langle a'|\right)|b'\rangle + \langle a'|\delta\left(|b'\rangle\right)$$
$$= \frac{i}{\hbar}\langle a'|(G_a - G_b)|b'\rangle = \frac{i}{\hbar}\langle a'|\delta W_{ab}|b'\rangle,$$

i.e., if the variations of the state functions are generated by a unitary operator, we can write:

$$\delta W_{ab} = G_a - G_b. \qquad (11.6)$$

(Note that $|a'\rangle$ means $|a', t\rangle$, so that $G_a$ depends, in general on $t$.)
Suppose we have an operator $O$ and calculate

$$\delta\langle a'|O|b'\rangle = \left(\delta\langle a'|\right)O|b'\rangle + \langle a'|O\delta\left(|b'\rangle\right)$$
$$= -\frac{i}{\hbar}\langle a'|(OG_b - G_a O)|b'\rangle.$$

Then we can define a corresponding infinitesimal operator, $O_{ab}$, so that

$$\delta\langle a'|O|b'\rangle \equiv \langle a'|\delta O_{ab}|b'\rangle$$

$$\text{and so infer} \quad \delta O_{ab} = -\frac{i}{\hbar}(OG_b - G_a O), \qquad (11.7)$$

where $\delta O_{ab}$ is the operator variation equivalent to the generations $G_{a,b}$ acting on the states.

All this is quite general. We now state and adopt Schwinger's Quantum Mechanical Action Principle: Of all possible variations $\delta W$ we consider only those obtained by the variation of a single operator $W$:

$$\delta W = \delta[W]. \qquad (11.8)$$

To obtain a simple, mechanical theory, we write $L(q(t), \dot{q}(t))$ and allow all possible variations of $W$. Then we expect

$$\delta \mathcal{W}_{12} = G(t_1) - G(t_2).$$

In particular , if  $t_1 = t_2 = t,$   $\delta O = -\dfrac{i}{\hbar}[O, G(t)].$          (11.9)

Here ends the introductory chapter on Schwinger's Action Principle in Quantum Mechanics.

Now we turn to quantum field theory.

# Reference

1.  Schwinger, J.: A note on the quantum dynamical principle. Phil. Mag. **44**(7), 1171 (1953)

# Chapter 12
# The Action Principle in Quantum Field Theory

This chapter is based in great part on the author's paper which appeared in 2015 in the European Physical Journal [1].Before we start with the most general covariant formulation of quantum field theory, we want to go back to Jordan, who contributed more than anybody else to the birth of quantum field theory in 1925, the year of the "Dreimännerarbeit" by Born, Heisenberg and Jordan [2]. Many experts in the field consider Dirac's paper of 1927 on quantum electrodynamics as the beginning of quantum field theory, but he did not do what everyone does today, i.e., use (as in classical field theory) the ordinary space–time coordinates $(\overrightarrow{r}, t)$ as argument of the quantum field operator. This is exactly what Euler did when he introduced the velocity vector field $\overrightarrow{v}(\overrightarrow{r}, t)$ in the eighteenth century for hydrodynamics.

At the end of a "Report on Quantum Electrodynamics" which appeared in "The Physicist's Conception of Nature," a book by Mehra (ed.) [3], published in 1973, one can find as part of the discussion in presence of a distinguished audience of physicists, the following remark by the famous Dutch physicist, van der Waerden:

> In talks on quantum field theory, we have heard from three different authors the statement that quantum field theory begins with Dirac's paper of 1927. This statement seems to be true in the minds of most physicists, in the minds of all physicists except one. Actually, quantum field theory begins with the paper of Jordan. You cannot write the history of quantum field theory neglecting him.

> There are two earlier papers on quantum field theory before Dirac's paper of 1927. Strangely enough, the papers themselves are well-known; i.e. the paper of Born and Jordan of 1925, on quantum mechanics, and the next paper of the year 1925 by Heisenberg, Born and Jordan, the famous three-men paper from which all the contemporary physicists learned quantum mechanics. At the end of these papers there is a section on quantum field theory, for which Jordan alone was responsible, not Born and not Heisenberg.

As a side note, let us not forget that the way we learn quantum mechanics (Q M.) today goes back to Born and Jordan's paper on Q.M., which appeared right after Heisenberg's famous article, both in the year 1925 [4]. Born asked Pauli to help him to make Heisenberg's discovery more understandable. Pauli refused, so Born turned

© The Author(s), under exclusive license to Springer Nature Switzerland AG 2021
W. Dittrich, *The Development of the Action Principle*,
SpringerBriefs in Physics,
https://doi.org/10.1007/978-3-030-69105-9_12

to Jordan and together, they were able to clean up Heisenberg's rather mysterious mathematics.

Nevertheless, we still talk about Heisenberg's matrix mechanics. As a matter of fact, Heisenberg did not even know what a matrix was. This we know from Born's autobiography, "My Life" [5], and from letters Born exchanged with Einstein [6]. Born immediately recognized the mathematics that was necessary to formulate the new physics. That Born could decipher Heisenberg's paper goes back to his student days in Breslau, now Poland, where he grew up and studied at the university there. Here he took courses in mathematics taught by Professor Rosanes, who, in 1901, lectured on matrices, those useless (for physicists) non-commuting objects.

Conclusion: The real birth of Q.M., which is still taught today, is due to Born and Jordan and consequently, we should speak of Born-Jordan Q.M.

Let us now begin with an explicit verification of Schwinger's action principle applied to a scalar field theory which uses an operator field $\Phi(\overrightarrow{r}, t)$—like Euler's velocity vector field $\overrightarrow{v}(\overrightarrow{r}, t)$ on a classical level!

The quantum field operator $\Phi(\overrightarrow{r}, t)$ is classically represented by a wave function $\varphi(\overrightarrow{r}, t)$ which describes a vibrating string with fixed end points. (As a side remark: the scalar field operator $\Phi(\overrightarrow{r}, t)$ or the Dirac spinor field operator $\Psi_\alpha(\overrightarrow{r}, t)$ are sometimes denoted as second quantized wave functions. This terminology is not only misleading; it is plain wrong. The Schrödinger or Dirac wave functions $(\varphi(\overrightarrow{r}, t), \psi_\alpha(\overrightarrow{r}, t))$ are *one-particle* wave functions, while $(\Phi(\overrightarrow{r}, t), \Psi_\alpha(\overrightarrow{r}, t))$ stand for the description of a many-particle system of Bose and Fermi particles. Hence, the Schrödinger wave function is not quantized a second time!

To demonstrate his approach to quantum field theory, Jordan uses a vibrating string with fixed endpoints at $x = 0$ and $x = l$. Let its linear mass density be $\sigma$ and its tension be $\tau$. Then the equation of motion for the vertical displacement $\Phi(x, t)$ at point $x$ and time $t$ is given by

$$\frac{1}{c^2} \frac{\partial^2 \Phi(x, t)}{\partial t^2} = \frac{\partial^2 \Phi(x, t)}{\partial x^2} \quad \text{with} \quad \Phi(0, t) = 0 = \Phi(l, t). \tag{12.1}$$

$c = \sqrt{\frac{\tau}{\sigma}}$ denotes the velocity in the string.

From now on, $\Phi(x, t)$ is considered a quantum field operator! As such, we need to formulate the action principle with the aid of a Lagrangian operator so that we can perform the variation

$$\mathcal{W}_{12} = \int_{t_2}^{t_1} dt \, L, \tag{12.2}$$

$$\delta \mathcal{W}_{12} = G[t_1] - G[t_2]$$

where $\quad L(t) = \int_0^l dx \, \mathcal{L}\left( \Phi(x, t), \frac{\partial \Phi(x, t)}{\partial t}, \frac{\partial \Phi(x, t)}{\partial x}; x, t \right) \quad$ (12.3)

The Lagrangian density $\mathcal{L}$ is itself a function of the field $\Phi(x,t)$. The role of the generalized velocity is taken over by $\frac{\partial \Phi(x,t)}{\partial t}$. The dependence on $\frac{\partial \Phi(x,t)}{\partial x}$ is new; it represents an "interaction" between the infinitesimally separated elements of the system. Note that $\mathcal{L}$ has the dimension of energy per unit length.

The generalized coordinate $\Phi(x,t)$ is varied according to

$$\Phi(x,t) \rightarrow \Phi(x,t) + \delta\Phi(x,t),$$

with the special boundary condition

$$\delta\Phi(x = 0, t) = 0 = \delta\Phi(x = l, t).$$

We will, however, allow for the time variation of the endpoints:

$$\delta\Phi(x, t_1) \neq 0 \neq \delta\Phi(x, t_2) \text{ for all } x.$$

Hamilton's principle is given by

$$\delta W_{12} = \int_{t_2}^{t_1} dt \left( L(t) \frac{d}{dt}(\delta t) + \delta L \right) \tag{12.4}$$

$$\text{with} \quad \mathcal{L} = \frac{\sigma}{2}\left(\frac{\partial \Phi}{\partial t}\right)^2 - \frac{\tau}{2}\left(\frac{\partial \Phi}{\partial x}\right)^2, \quad \sigma, \tau = \text{const.}$$

$$= \mathcal{T} - \mathcal{V}. \tag{12.5}$$

Using

$$\delta L = \frac{\partial \mathcal{L}}{\partial t}\delta t + \frac{\partial \mathcal{L}}{\partial x}\delta x + \frac{\partial \mathcal{L}}{\partial \Phi}\delta\Phi + \frac{\partial \mathcal{L}}{\partial\left(\frac{\partial \Phi}{\partial t}\right)}\delta\left(\frac{\partial \Phi}{\partial t}\right) + \frac{\partial \mathcal{L}}{\partial\left(\frac{\partial \Phi}{\partial x}\right)}\delta\left(\frac{\partial \Phi}{\partial x}\right)$$

with

$$\frac{\partial \mathcal{L}}{\partial \Phi} = 0, \quad \frac{\partial \mathcal{L}}{\partial\left(\frac{\partial \Phi}{\partial t}\right)} = \sigma\left(\frac{\partial \Phi}{\partial t}\right), \quad \frac{\partial \mathcal{L}}{\partial\left(\frac{\partial \Phi}{\partial x}\right)} = -\tau\left(\frac{\partial \Phi}{\partial x}\right),$$

we obtain

$$\delta L = \sigma \frac{\partial \Phi}{\partial t}\frac{\partial}{\partial t}(\delta\Phi) - \sigma\left(\frac{\partial \Phi}{\partial t}\right)^2 \frac{\partial}{\partial t}(\delta t) - \tau \frac{\partial \Phi}{\partial x}\frac{\partial}{\partial x}(\delta\Phi).$$

After a few more steps we finally end up with the Lagrangian formulation:

$$\delta W_{12} = \int_{t_2}^{t_1} dt \int_0^l dx \left\{ \frac{\partial}{\partial t} \left[ \sigma \left( \frac{\partial \Phi}{\partial t} \right) \delta \Phi - \left( \frac{\sigma}{2} \left( \frac{\partial \Phi}{\partial t} \right)^2 + \frac{\tau}{2} \left( \frac{\partial \Phi}{\partial x} \right)^2 \right) dt \right] \right.$$

$$\left. - \left( \sigma \left( \frac{\partial^2 \Phi}{\partial t^2} - \tau \frac{\partial^2 \Phi}{\partial x^2} \right) \delta \Phi + \delta t \frac{\partial}{\partial t} \left[ \frac{\sigma}{2} \left( \frac{\partial \Phi}{\partial t} \right)^2 + \frac{\tau}{2} \left( \frac{\partial \Phi}{\partial x} \right)^2 \right] \right) \right\}. \qquad (12.6)$$

From here we can identify $(\delta W_{12} = 0)$ :

$\delta \Phi : \sigma \left( \frac{\partial \Phi}{\partial t^2} \right) - \tau \frac{\partial^2 \Phi}{\partial x^2} = 0$ equation of motion, i.e., one-dimensional string equation

$$\delta t : \frac{\partial}{\partial t} \left[ \frac{\sigma}{2} \left( \frac{\partial \Phi}{\partial t} \right)^2 + \frac{\tau}{2} \left( \frac{\partial \Phi}{\partial x} \right)^2 \right] = 0 \text{ energy conservation} \qquad (12.7)$$

or

$$\frac{\partial}{\partial t} (\mathcal{T} + \mathcal{V} = \varepsilon) = 0. \qquad (12.8)$$

Surface term

$$G = \int_0^l dx \left[ \sigma \left( \frac{\partial \Phi}{\partial t} \right) \delta \Phi - \varepsilon \delta t \right] \qquad (12.9)$$

with

$$\Pi = \sigma \left( \frac{\partial \Phi}{\partial t} \right), \quad \text{momentum density} \qquad (12.10)$$

and

$$\varepsilon = \frac{\sigma}{2} \left( \frac{\partial \Phi}{\partial t} \right)^2 + \frac{\tau}{2} \left( \frac{\partial \Phi}{\partial x} \right)^2, \quad \text{total energy density}$$

$$E = \int_0^l dx \varepsilon. \qquad (12.11)$$

The momentum density conjugate to $\Phi(x, t)$ is defined as

$$\Pi(x, t) = \frac{\partial \mathcal{L}}{\partial \left( \frac{\partial \Phi}{\partial t} \right)} \left( = \sigma \frac{\partial \Phi}{\partial t} \right) \text{ for the string.} \qquad (12.12)$$

The Hamiltonian density is in general given by

$$\mathcal{H}\left(\Phi, \Pi, \frac{\partial \Phi}{\partial x}; x, t\right) = \Pi \frac{\partial \Phi}{\partial t} - \mathcal{L}.$$

For our one-dimensional string this yields with (12.8) and (12.12):

$$\mathcal{H} = \sigma \frac{\partial \Phi}{\partial t} \frac{\partial \Phi}{\partial t} - \left(\frac{\sigma}{2}\left(\frac{\partial \Phi}{\partial t}\right)^2 - \frac{\tau}{2}\left(\frac{\partial \Phi}{\partial x}\right)^2\right) = \frac{\sigma}{2}\left(\frac{\partial \Phi}{\partial t}\right)^2 + \frac{\tau}{2}\left(\frac{\partial \Phi}{\partial x}\right)^2$$

$$= \frac{1}{2\sigma}(\Pi(x,t))^2 + \frac{\tau}{2}\left(\frac{\partial \Phi}{\partial x}\right)^2 = T + V$$

and therefore

$$H(t) = \int_0^l dx \, \mathcal{H}(x, t). \tag{12.13}$$

In the Hamiltonian formulation we obtain:

$$\delta W_{12} = \int_2^1 dt \int_0^l dx \left\{ \frac{\partial}{\partial t}[\Pi \delta \Phi - \mathcal{H} \, \delta t] \right.$$

$$+ \delta \Phi \left( -\frac{\partial \Pi}{\partial t} - \frac{\partial \mathcal{H}}{\partial \Phi} + \frac{\partial}{\partial x} \frac{\partial \mathcal{H}}{\partial\left(\frac{\partial \Phi}{\partial x}\right)} \right)$$

$$\left. + \delta \Pi \left( \frac{\partial \Phi}{\partial t} - \frac{\partial \mathcal{H}}{\partial \Pi} \right) + \delta t \frac{\partial}{\partial t} \mathcal{H} \right) \tag{12.14}$$

so that $\delta W_{12} = 0$ implies

$$\delta \Phi : \frac{\partial \Pi}{\partial t} = -\frac{\partial \mathcal{H}}{\partial \Phi} + \frac{\partial}{\partial x} \frac{\partial \mathcal{H}}{\partial\left(\frac{\partial \Phi}{\partial x}\right)},$$

$$\delta \Pi : \frac{\partial \Phi}{\partial t} = \frac{\partial \mathcal{H}}{\partial \Pi},$$

$$\delta t : \frac{\partial \mathcal{H}}{\partial t} = 0. \tag{12.15}$$

Surface term:

$$G = \int_0^l dx \, (\Pi(x,t)\delta \Phi(x,t) - \mathcal{H}\delta t), \tag{12.16}$$

$$\mathcal{H} = \frac{1}{2\sigma}(\Pi(x,t))^2 + \frac{\tau}{2}\left(\frac{\partial \Phi}{\partial x}\right)^2. \tag{12.17}$$

Now, the quantum field operators $\Phi(x, t)$ and $\Pi(x, t)$ have to satisfy certain equal-time commutation relations. To derive these relations, I will make later on the following fundamental operator statement of the quantum action principle: Suppose we have an operator $O = (\Phi(t))$ and we consider a variation corresponding to a change of the parametric variable $t$; then the infinitesimal change of the operator $O$ is given by

$$\delta O = -\frac{i}{\hbar}[O, G].  \tag{12.18}$$

Let at a given time ($\delta t = 0$) the variation $\delta \Phi(x, t) \neq 0$; then from the generator (12.16) we obtain

$$G(t) = \int_0^l dx' \Pi(x', t) \delta \Phi(x', t).$$

Now, using (12.18) with $O = \Phi(x, t)$, we have:

$$\delta \Phi(x, t) = -\frac{i}{\hbar}\left[\Phi(x, t), \int_0^l dx' \Pi(x', t) \delta \Phi(x', t)\right],$$

or

$$\left[\Phi(x, t), \Pi(x', t) = i\hbar\delta(x - x').\right]  \tag{12.19}$$

Here is a list of Jordan's equal-time commutation relations in the Heisenberg picture:

$$\left[\Phi(x, t), \Phi(x', t) = 0\right], \quad \left[q_i, q_j\right] = 0,$$

$$\left[\Pi(x, t), \Pi(x', t) = 0\right], \quad \left[p_i, p_j\right] = 0,$$

$$\left[\Phi(x, t), \Pi(x', t)\right] = i\hbar\delta(x - x') \quad \left[q_i, p_j\right] = i\hbar\delta_{ij}.  \tag{12.20}$$

Using the Born-Jordan commutation relations in dimensionless form we have

$$\left[a_n, a_m^\dagger\right] = \delta_{nm}, \qquad [q_n(t), p_m(t)] = i\hbar\delta_{nm}.  \tag{12.21}$$

Then the transition from the quantum matter-field to the total energy operator of the string is given by

$$H = \frac{1}{2} \int_0^l dx \left[ \sigma \left( \frac{\partial \Phi}{\partial t} \right)^2 + \tau \left( \frac{\partial \Phi}{\partial x} \right)^2 \right] = \frac{1}{2} \sum_{n=1}^{\infty} \hbar \omega_n (a_n^\dagger a_n + a_n a_n^\dagger)$$

$$= \sum_{n=1}^{\infty} \hbar \omega_n \left( a_n^\dagger a_n + \frac{1}{2} \right). \tag{12.22}$$

With the following replacements of the operators by c-numbers,

$$a_n \rightarrow \left( \frac{\omega_n}{2\hbar} \right)^{\frac{1}{2}} c_n e^{-i\phi_n}, \, a_n^\dagger \rightarrow \left( \frac{\omega_n}{2\hbar} \right)^{\frac{1}{2}} c_n e^{i\phi_n},$$

we get the classical expression, $E = T + U = \frac{1}{2} \sum_{n=1}^{\infty} \omega_n^2 c_n^2$,

for the total energy of the string. Hence, the real classical dynamical variables take the form:

$$\left( \frac{\hbar}{\omega_n} \right)^{\frac{1}{2}} q_n(t) \rightarrow q_n^{cl}(t) = c_n \cos(\omega_n t + \phi_n),$$

$$\left( \frac{\hbar}{\omega_n} \right)^{\frac{1}{2}} p_n(t) \rightarrow p_n^{cl}(t) = -\omega_n c_n \sin(\omega_n t + \phi_n) = \dot{q}_n^{cl}(t).$$

Now we are in a position to present a short introduction to the calculation of the mean-square energy fluctuation for the quantized string field.

Let us first calculate the energy $E_{a,\omega}$ of a piece of string of length $a$ in the narrow frequency range , $\omega + \Delta\omega$:

$$E(t) = \frac{1}{2} \int_0^a dx \left[ \sigma \left( \frac{\partial \Phi}{\partial t} \right)^2 + \tau \left( \frac{\partial \Phi}{\partial x} \right)^2 \right].$$

One obtains with $H$ from (12.22):

$$E_{a,\omega}(t) = \frac{a}{l} H + \frac{1}{2l} \sum_{n \neq m} \hbar \sqrt{\omega_n \omega_m} \left[ q_n q_m K'_{nm} + p_n p_m K_{nm} \right]$$

$$+ \sum_n \hbar \omega_n (q_n^2 - p_n^2) \sin(2k_n a) \frac{1}{2k_n}, \quad \omega_n = ck_n = \left( \frac{c\pi}{l} \right) n, \quad \lambda_n = \frac{2l}{n}. \tag{12.23}$$

$$K'_{nm} = \frac{\sin(k_n - k_m)a}{k_n - k_m} + \frac{\sin(k_n + k_m)a}{k_n + k_m},$$

$$K_{nm} = \frac{\sin(k_n - k_m)a}{k_n - k_m} - \frac{\sin(k_n + k_m)a}{k_n + k_m}. \tag{12.24}$$

Since we will be interested in the fluctuations in a narrow spectral range with.

$\lambda_n = \frac{2l}{n}$ much smaller than both $l$ and $a$, the final term in (12.23) can be neglected due to fast oscillations, so the deviation of the energy from the diagonal part, $\frac{a}{l} H$, in (12.23) is given by

$$\Delta E_{a,\omega} = \frac{1}{2l} \sum_{n \neq m} \hbar \sqrt{\omega_n \omega_m} [q_n q_m K'_{nm} + p_n p_m K_{nm}].$$ (12.25)

From here on, the reader can follow Jordan's explicit calculations outlined in the Drei-Männer-Arbeit by Born, Heisenberg and Jordan. Furthermore, it is rewarding to read up on the string field history in Heisenberg's Chicago lecture of 1927 and Jordan's conference contribution in Charkow in May of 1929.

The result of Jordan's work is:

$$\overline{\Delta E_{a,\nu}^2} = \frac{c}{2a\,\Delta\nu} \overline{E}_{a,\nu}^2 + h\nu \overline{E}_{a,\nu}$$ (12.26)

in complete analogy with Einstein's mean square energy fluctuation formula for black body radiation:

$$\overline{\Delta E_{V,\nu}^2} = \frac{c^3}{8\pi \nu^2 V \Delta\nu} \overline{E}_{V,\nu}^2 + h\nu \overline{E}_{V,\nu}.$$ (12.27)

In either case, the origin of the particle-like fluctuation, second part in (12.26), is the presence of the zero-point energy term due to the non-commutativity of the quantized amplitudes $q_n(t)$ and $p_m(t)$. The appearance of this particle-like fluctuation is a kinematic effect which arises from the quantized nature of the string (photon) field. Jordan himself considered his fluctuation calculation as "almost the most important contribution I made to quantum mechanics". In fact, it was the beginning of quantum field theory. However, very shortly thereafter, Dirac started his first attempt to formulate quantum electrodynamics. Nevertheless, Jordan introduced for the first time the quantum field operator as equivalent description of many particle wave functions. This is, until the present day, the accepted language for treating relativistic particle processes – with all its inherent problems that stem from products of relativistic local quantum field operators.

# References

1. Dittrich, W.: The cofounder of quantum field theory: Pascal Jordan. The Eur. Phy. J. **H 40**, 241–260 (2015)
2. Born, M., Heisenberg, W., Jordan, P.: Zur Quantenmechanik II. Z. Phys. **30**, 558 (1925)
3. Mehra, J. (Ed.): The Physicist's Conception of Nature. Reidel Publishing Company, Dordrecht-Holland (1973)
4. Born, M., Jordan, P.: Zur Quantenmechanik. Z. Phys. **34**, 858 (1925)

5. Born, M.: My Life: Recollections of a Nobel Laureate. Scribner, New York (1978)
6. Einstein, A., Born, M.: Born Einstein Letters, 1916-1955: Friendship, Politics and Physics in Uncertain Times. MacMillan Ltd, London (1971)

# Chapter 13
# Quantum Field Theory on Space-Like Hypersurfaces

In this chapter we want to demonstrate that the equations of motion and commutation relations can be generalized to replace the time $t$ to a space-like hypersurface. Our procedure is based again on the action principle.

Our notation: $g = diag\ (1,\ 1,\ 1, -1)$, $\hbar = 1 = c$, "dimensionless" action $[\mathcal{L}] = L^{-4}$.

The action operator is now given its most general form:

$$W\{[\Phi(x)]; \sigma_1, \sigma_2\} = \int_{\sigma_1}^{\sigma_2} d^4x \mathcal{L}\big(\Phi, \partial_\mu \Phi\big),\ d^4x = dx^0 d^3\vec{x} \tag{13.1}$$

$\Phi \equiv \varphi_\alpha$, collection of local fields, scalars, spinors, etc.
As before, the action principle takes the form

$$\delta W = G(\sigma_2) - G(\sigma_1), \tag{13.2}$$

where now the flat t-plane has been replaced by $\sigma(x)$ a space-like hypersurface $(x - x')^2 > 0$ (Fig. 13.1).

Now we turn our attention to the response of $W$ under the change

$$x'^\mu = x^\mu + \delta x^\mu(x). \tag{13.3}$$

The change of the volume element can be shown to be

$$\delta\big(d^4x\big) = d^4x\big(\partial_\mu \delta x^\mu\big). \tag{13.4}$$

The variation of the action is then

$$\delta W = \delta \int_{\sigma_1}^{\sigma_2} d^4x \mathcal{L}\big(\Phi, \partial_\mu \Phi\big)$$

© The Author(s), under exclusive license to Springer Nature Switzerland AG 2021
W. Dittrich, *The Development of the Action Principle*,
SpringerBriefs in Physics,
https://doi.org/10.1007/978-3-030-69105-9_13

**Fig. 13.1** Space-like
hypersurfaces

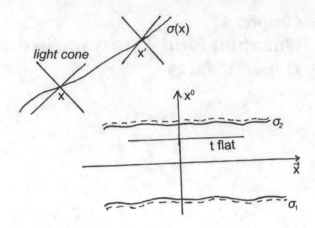

$$= \int_{\sigma_1}^{\sigma_2} \left[ \delta\left(d^4x\right)\mathcal{L} + d^4x\delta\mathcal{L} \right]$$

$$= \int_{\sigma_1}^{\sigma_2} d^4x \left[ \left(\partial_\mu \delta x^\mu\right)\mathcal{L} + \delta\mathcal{L} \right] \tag{13.5}$$

Now apply

$$\delta f = f'\left(x'\right) - f(x) = f'(x + \delta x) - f(x)$$
$$= \underline{(f'(x) - f(x))} + \delta x^\mu \underline{\partial_\mu f'} + \mathcal{O}\left(\delta x^2\right)$$
$$=: \delta_0 f, \text{ fctl. change } \mathcal{O}\left(\delta\left(x^0\right)\right) \to \partial_\mu f \text{ local change}$$
$$> \delta f = \delta_0 f + \delta x^\mu \partial_\mu f. \tag{13.6}$$

In general,

$$\delta \cdot / \cdot = \delta_0 \cdot / \cdot + \delta x^\mu \partial_\mu \cdot / \cdot \tag{13.7}$$

We need in (13.5)

$$\delta\mathcal{L} = \delta_0\mathcal{L} + \delta x^\mu \partial_\mu \mathcal{L}$$
$$= \frac{\partial\mathcal{L}}{\partial\Phi}\delta_0\Phi + \frac{\partial\mathcal{L}}{\partial[\partial_\mu\Phi]}\delta_0(\partial_\mu\Phi) + \delta x^\mu \partial_\mu \mathcal{L}$$

$\delta_0 \equiv$ functional change, therefore in 2nd term

$$\delta_0(\partial_\mu\Phi) = \left[\delta_0, \partial_\mu\right]\Phi + \partial_\mu(\delta_0\Phi) = \partial_\mu(\delta_0\Phi)$$
$$= 0$$

$$> \delta\mathcal{L} = \frac{\partial\mathcal{L}}{\partial\Phi}\delta_0\Phi + \frac{\partial\mathcal{L}}{\partial[\partial_\mu\Phi]}\partial_\mu(\delta_0\Phi) + \delta x^\mu\partial_\mu\mathcal{L}$$

$$= \partial_\mu\left(\frac{\partial\mathcal{L}}{\partial[\partial_\mu\Phi]}\delta_0\Phi\right) - \partial_\mu\left(\frac{\partial\mathcal{L}}{\partial[\partial_\mu\Phi]}\right)\delta_0\Phi$$

$$= \delta x^\mu\partial_\mu\mathcal{L} + \left[\frac{\partial\mathcal{L}}{\partial\Phi} - \partial_\mu\frac{\partial\mathcal{L}}{\partial[\partial_\mu\Phi]}\right]\delta_0\Phi + \partial_\mu\left(\frac{\partial\mathcal{L}}{\partial[\partial_\mu\Phi]}\delta_0\Phi\right).$$

When substituted in (13.2), (13.5), the second term yields the Euler–Lagrange equations, i.e., the equations of motion:

$$\boxed{\frac{\partial\mathcal{L}}{\partial\Phi} - \partial_\mu\frac{\partial\mathcal{L}}{\partial[\partial_\mu\Phi]} = 0.} \qquad (13.8)$$

So far we have ("on shell"):

$$\delta W = \int_{\sigma_1}^{\sigma_2} d^4x\left[(\partial_\mu\delta x^\mu)\mathcal{L} + \delta x^\mu\partial_\mu\mathcal{L} + \partial_\mu\left(\frac{\partial\mathcal{L}}{\partial[\partial_\mu\Phi]}\delta_0\Phi\right)\right]$$

$$= \partial_\mu(\mathcal{L}\delta x^\mu)$$

$$= \int_{\sigma_1}^{\sigma_2} d^4x\,\partial_\mu\left[\mathcal{L}\delta x^\mu + \frac{\partial\mathcal{L}}{\partial[\partial_\mu\Phi]}\delta_0\Phi\right]$$

$$(7) = (\delta - \delta x^\nu\partial_\nu)\Phi$$

$$= \int_{\sigma_1}^{\sigma_2} d^4x\,\partial_\mu\left[\frac{\partial\mathcal{L}}{\partial[\partial_\mu\Phi]}\delta\Phi - \delta x^\nu\left(\frac{\partial\mathcal{L}}{\partial[\partial_\mu\Phi]}\partial_\nu\Phi - \mathcal{L}g^\mu_\nu\right)\right]$$

Action Principle $= G(\sigma_2) - G(\sigma_1)$,
where $G(\sigma)$ is given by (Gauss' theorem)

$$\boxed{G(\sigma) = \int_\sigma d\sigma_\mu\{\Pi^\mu\delta\Phi - (\Pi^\mu\partial^\nu\Phi - g^{\mu\nu}\mathcal{L})\delta x_\nu\}} \qquad (13.9)$$

and

$$\Pi^\mu := \frac{\partial\mathcal{L}}{\partial[\partial_\mu\Phi]} \quad \text{"can. momentum"} \qquad (13.10)$$

$$T^{\mu\nu}(x) = \Pi^\mu(x)\partial^\nu\Phi(x) - g^{\mu\nu}\mathcal{L} \quad \text{can. energy} - \text{mom. tensor}$$

$$\left(T^{00} = T_{00} = \Pi^0(x)\partial^0\Psi(x) - \mathcal{L}(x) = \mathcal{H}(x)\right)$$

$$H = \int d^3\vec{x}\,\mathcal{H}(x) = \int d^3\vec{x}\,T^{00} = P^0 = -P_0$$

$$\delta x_0 = -\delta x^0 > T^{00}(-)\delta x^0 = -H\delta t \tag{13.11}$$

$\int\limits_{\sigma_1}^{\sigma_2} d^4x \partial_\mu F^\mu = \int\limits_{\sigma_1}^{\sigma_2} d\sigma_\mu F^\mu \to \int\limits_{t_1}^{t_2} d\sigma_0 F^0$, to be compared with.

Gauss flat $\sigma$

$$\int\limits_{\sigma_1}^{\sigma_2} d^4x \partial_\mu F^\mu = \int\limits_{\sigma_1}^{\sigma_2} d^4x (\partial_0 F^0 + \vec{\nabla} \cdot \vec{F}) \to \int d^3\vec{x} F^0 \int d^3\vec{x} F^0 \big]_{t_1}^{t_2}.$$

Gauss.

i.e., $d\sigma_0$ at a given time $\equiv d^3 \vec{x}$

Operator statement of action principle.:

$$\hbar = 1 \delta \mathcal{O} = \frac{1}{i}[\mathcal{O}, G] \tag{13.12}$$

Now choose $\delta x^\nu = 0$ in (13.9). Then we obtain

$$G(\sigma) = \int\limits_\sigma d\sigma_\mu \frac{\partial \mathcal{L}}{\partial[\partial_\mu \Phi]} \delta \Phi$$

$$\equiv \Pi^\mu \tag{13.13}$$

For simplicity consider $\sigma \to t$; then

$$\Pi \equiv \Pi^0 G(\sigma) \to G(t) = \int d^3\vec{x}' \Pi(\vec{x}', t)\delta\Phi(\vec{x}', t) \tag{13.14}$$

Using (13.12) with $\mathcal{O} := \Phi(\vec{x}, t)$, we have

$$\delta\Phi(\vec{x}, t) = \frac{1}{i}\left[\Phi(\vec{x}, t), \int d^3\vec{x}' \Pi(\vec{x}', t)\delta\Phi(\vec{x}', t)\right] \tag{13.15}$$

$$[A, BC] = [A, B]C + B[A, C] = \frac{1}{i}\int d^3\vec{x}'([\Phi(\vec{x}, t), \Pi(\vec{x}', t)]\delta\Phi(\vec{x}', t)$$

$$+ \Pi(\vec{x}', t)[\Phi(\vec{x}, t), \delta\Phi(\vec{x}', t)]$$

$$\boxed{> [\Phi(\vec{x}, t), \Pi(\vec{x}', t)] = i\delta(\vec{x} - \vec{x}')} \quad \text{E.T.C.R.}$$

$$\tag{13.16}$$

$$[\Phi(\vec{x}, t), \delta\Phi(\vec{x}', t)] = 0 \quad (= [\delta\Phi(\vec{x}, t), \Phi(\vec{x}', t)])$$

$$> \quad 0 = [\delta\Phi(\vec{x}, t), \Phi(\vec{x}', t)] + [\Phi(\vec{x}, t), \delta\Phi(\vec{x}', t)]$$

$$\equiv \delta[\Phi(\vec{x}, t), \Phi(\vec{x}', t)]$$

$$\boxed{[\Phi(\vec{x}, t), \Phi(\vec{x}', t)] = 0} \tag{13.17}$$

Similarly, from

$$\delta \Pi\left(\vec{x}, t\right) = \frac{1}{i}\left[\Pi\left(\vec{x}, t\right), \int d^3 \vec{x}' \, \Pi\left(\vec{x}', t\right) \delta \Phi\left(\vec{x}', t\right)\right],$$

we obtain

$$\boxed{\left[\Pi(\vec{x}, t), \Pi(\vec{x}', t)\right] = 0} \tag{13.18}$$

In general, for arbitrary space-like separations, $(x - x')^2 > 0$,

$$\left[\Phi_\alpha(x), \Pi_\beta(x')\right] = i\delta_{\alpha\beta}\delta(x - x')$$
$$\left[\Phi_\alpha(x), \Phi_\beta(x')\right] = 0 = \left[\Pi_\alpha(x), \Pi_\beta(x')\right] \tag{13.19}$$

All this holds for Bose fields.

If we are dealing with fermions, we want ETACR's. They may be included if we consider $\delta\psi$ as anticommuting with all operators. Then

$$\delta\psi = \frac{1}{i}\left[\psi, \int d^3\vec{x}' \Pi' \delta\psi'\right] = \frac{1}{i} \int d^3\vec{x}' \left(\psi \Pi' \delta\psi' - \Pi' \underline{\delta\psi'\psi}\right)$$
$$- \psi\delta\psi'$$
$$= \frac{1}{i} \int d^3\vec{x}' \left(\psi \Pi' + \Pi'\psi\right)\delta\psi'$$

Fermi–Dirac

$$> \quad \left\{\psi_\alpha(\vec{x}, t), \Pi_\beta(\vec{x}', t)\right\} = i\delta_{\alpha\beta}\delta(\vec{x} - \vec{x}')$$
$$\left\{\psi_\alpha(\vec{x}, t), \psi_\beta(\vec{x}', t)\right\} = 0 = \left\{\Pi_\alpha(\vec{x}, t), \Pi_\beta(\vec{x}', t)\right\}. \tag{13.20}$$

# Chapter 14
# Lagrangian Formulation of Gauge Theories

Let us start with the well-known Abelian gauge theory quantum electrodynamics (QED). The non-interacting particles with mass $m$ are described by the free-field Lagrangian:

$$\mathcal{L} = \overline{\psi}(x)(i\gamma \cdot \partial - m)\psi(x). \tag{14.1}$$

The Lagrangian is invariant under the global $U(1)$ phase transformation:

$$\psi(x) \to \psi'(x) = e^{i\Lambda}\psi(x)$$
$$\overline{\psi}(x) \to \overline{\psi}'(x) = \overline{\psi}(x)e^{-i\Lambda}, \tag{14.2}$$

with $\Lambda = \text{const.}$, $\Lambda \in \mathbb{R}$. This is an algebraic symmetry and is called gauge transformation of the first kind.

Now we want to make this symmetry a local symmetry, i.e., we want to gauge the $U(1)$ symmetry by replacing the constant $\Lambda$ by $\Lambda(x)$, where $\Lambda(x)$ is as arbitrary function of space–time position. Consequently, (14.2) is turned into a local phase transformation:

$$\psi(x) \to \psi'(x) = e^{i\Lambda(x)}\psi(x)$$
$$\overline{\psi}(x) \to \overline{\psi}'(x) = \overline{\psi}(x)e^{-i\Lambda(x)} \tag{14.3}$$

This is called a gauge transformation of the second kind. It will become clear in a moment that this is a dynamical, or experimental, symmetry.

Our goal is to construct an interacting theory which will be invariant under the gauge transformation of the second kind. Obviously, $\mathcal{L}$ in (14.1) is not invariant under (14.3), because we pick up a $\partial_\mu \Lambda(x)$ term. As is familiar, we can obtain a

© The Author(s), under exclusive license to Springer Nature Switzerland AG 2021
W. Dittrich, *The Development of the Action Principle*,
SpringerBriefs in Physics,
https://doi.org/10.1007/978-3-030-69105-9_14

total invariant Lagrangian if we introduce a vector field $A_\mu(x)$, which couples to the spinor field $\psi$ according to.

$$\mathcal{L} = \overline{\psi} i \gamma (\partial - ieA)\psi - \overline{\psi} m \psi - \frac{1}{4} F_{\mu\nu} F^{\mu\nu}, \tag{14.4}$$

$$F_{\mu\nu} = \partial_\mu A_\nu - \partial_\nu A_\mu. \tag{14.5}$$

The Lagrangian (14.4) is then locally invariant under the gauge transformation of the second kind (14.3) and $A_\mu(x)$ is shifted by the transformation

$$A_\mu(x) A'_\mu(x) = A_\mu(x) + \partial_\mu \Lambda(x). \tag{14.6}$$

(The same $\Lambda$ as in 14.3.)

The proof of this statement is trivial and will not be presented here; put $\Lambda(x) e \Lambda(x)$ into (14.3).

The special combination

$$D_\mu := \partial_\mu - ieA_\mu \tag{14.7}$$

is called the gauge-covariant derivative.

Taken together, the Lagrangian (14.4) is seen to be formed out of the various fields $\psi$, their gauge-covariant derivatives and the gauge-invariant kinetic term of the photo field, $-\frac{1}{4}F^2$, constructed from the gauge field derivative according to (14.5).

A photon mass term $A_\mu A^\mu$ is not permitted, since it is not gauge invariant. Also note that under gauge transformation of the second kind, $D_\mu \psi$ behaves like $\psi$, Eq. (14.3), namely

$$D_\mu \psi \left[ D_\mu \psi(x) \right]' = e^{ie\Lambda(x)} D_\mu \psi(x). \tag{14.8}$$

By direct substitution of (14.6) into (14.4), it is easy to see that $F$ is, in fact, gauge invariant. It can, however, also be seen as follows: the antisymmetric field strength tensor $F_{\mu\nu}$ is related to the commutator of two covariant derivatives as

$$[D_\mu, D_\nu]\psi = -ieF_{\mu\nu}\psi. \tag{14.9}$$

From (14.8) it follow that:

$$\left( [D_\mu, D_\nu]\psi \right)' = e^{ie\Lambda(x)} \left( [D_\mu, D_\nu]\psi \right) \tag{14.10}$$

or

$$F'_{\mu\nu}\psi' = e^{ie\Lambda(x)} F_{\mu\nu}\psi = F_{\mu\nu}\psi', \tag{14.11}$$

which means

$$F'_{\mu\nu} = F_{\mu\nu}. \tag{14.12}$$

Thus, any Lagrangian constructed from $\psi$, $D_\mu\psi$ and $F_{\mu\nu}$ is invariant under gauge transformation of the second kind if it is invariant under gauge transformation of the first kind.

$$\mathcal{L}(\psi, \partial_\mu\psi)\mathcal{L}_{matter}(\psi, D_\mu\psi) - \frac{1}{4}F_{\mu\nu}F^{\mu\nu} \tag{14.13}$$

$$\overline{\psi}(i\gamma \cdot \partial - m)\psi\overline{\psi}(i\gamma \cdot D - m)\psi - \frac{1}{4}F_{\mu\nu}F^{\mu\nu}. \tag{14.14}$$

As is well known, conserved (and partially conserved) currents are related to symmetry properties. In order to see this relationship, consider a field theory described by the Lagrangian.

$$\mathcal{L}(x) = \mathcal{L}(\psi_i(x), \partial_\mu\psi_i(x)), i = 1, 2, \ldots n, \tag{14.15}$$

where the fields transform to some symmetry group $G$ whose Lie algebra is given by

$$[T^i, T^j] = i f_{ijk} T^k \tag{14.16}$$

where the $T^l$ are the generators of the unitary group. The $f_{ijk}$ are the structure constants of the (compact) Lie group:

$$f_{ijl}f_{lkm} + f_{jkl}f_{lim} + f_{kil}f_{ljm} = 0.$$

Now let us perform an infinitesimal linear local phase transformation on the fields:

$$\psi_i(x)\psi'_i(x) = \psi_i(x) + i\Lambda^l(x)T^l_{ij}\psi_j(x) \tag{14.17}$$

$$\partial_\mu\psi_i(x)\partial_\mu\psi'_i(x) = \partial_\mu\psi_i(x) + i\left(\partial_\mu\Lambda^l(x)\right)T^l_{ij}\psi_j(x) + i\Lambda^l(x)T^l_{ij}(\partial_\mu\psi_j(x)\right). \tag{14.18}$$

The $\Lambda^l(x)$ are infinitesimal space–time functions.
To first order in $\Lambda^l(x)$, the change in $\mathcal{L}$ is given by

$$\delta\mathcal{L} = \frac{\partial\mathcal{L}}{\partial\psi_i}\delta\psi_i + \frac{\partial\mathcal{L}}{\partial(\partial_\mu\psi_i)}\delta(\partial_\mu\psi_i). \tag{14.19}$$

Then, with $\delta\psi_i = \psi'_i(x) - \psi_i(x)$, $\delta(\partial_\mu\psi_i) = \partial_\mu\psi'_i(x) - \partial_\mu\psi_i(x)$, which we insert in (14.19), and obtain after a few steps which include

$$\frac{\delta\mathcal{L}}{\delta(\partial_\mu\Lambda^l)} = i\frac{\partial\mathcal{L}}{\partial(\partial_\mu\psi_i)}T^l_{ij}\psi_j,$$
(14.20)

and the Euler equation,

$$\partial_\mu\left(\frac{\partial\mathcal{L}}{\partial(\partial_\mu\psi_i)}\right) = \frac{\partial\mathcal{L}}{\partial\psi_i},$$
(14.21)

the (Gell-Mann-Levy) equation:

$$\partial_\mu\frac{\delta\mathcal{L}}{\delta(\partial_\mu\Lambda^l)} = \frac{\delta\mathcal{L}}{\delta\Lambda^l}.$$
(14.22)

This equation looks like an Euler–Lagrange equation which it isn't, because $\Lambda^l(x)$ is a gauge function and not a dynamical variable.

If we define $J^l_\mu(x)$ to be the current associated with the phase transformation (14.17), i.e.,

$$J^l_\mu(x) = -\frac{\delta\mathcal{L}}{\delta(\partial^\mu\Lambda^l)} = -i\frac{\partial\mathcal{L}}{\partial(\partial^\mu\psi_i)}T^l_{ij}\psi_j,$$
(14.23)

then its divergence is, according to (14.22),

$$\partial^\mu J^l_\mu(x) = -\partial^\mu\frac{\delta\mathcal{L}}{\delta(\partial_\mu\Lambda^l)} = -\frac{\delta\mathcal{L}}{\delta\Lambda^l}$$

or

$$\partial^\mu J^l_\mu(x) = -\frac{\delta\mathcal{L}}{\delta\Lambda^l}.$$
(14.24)

Therefore, we can compute the divergence of $J^l_\mu$ directly from $\mathcal{L}(\Lambda)$ without use of the equations of motion. Moreover, if $\frac{\delta\mathcal{L}(\Lambda)}{\delta\Lambda^l} = 0$, then the current is conserved.

So far, we have found the response of $\mathcal{L}$ under infinitesimal phase transformation to be

$$\delta\mathcal{L} = \frac{\delta\mathcal{L}}{\delta\Lambda^l}\Lambda^l(x) + \frac{\delta\mathcal{L}}{\delta(\partial_\mu\Lambda^l)}(\partial_\mu\Lambda^l(x))$$
$$= -\left(\partial_\mu J^{\mu l}(x)\right)\Lambda^l(x) - J^{\mu l}(\partial_\mu\Lambda^l(x))$$
(14.25)

$$= -\partial_\mu \left( J^{\mu l}(x) \Lambda^l(x) \right).$$  (14.26)

In particular, if $\Lambda^l(x) = const.$, i.e.,

$$\delta \mathcal{L} = -\left( \partial_\mu J^{\mu l}(x) \right) \Lambda^l(x),$$  (14.27)

we learn that for constant infinitesimal phase transformation, $\delta \mathcal{L} = 0$ implies $\partial_\mu J^{\mu l}(x) = 0$. Thus, with any constant infinitesimal phase transformation which leaves the Lagrangian invariant, there exists a conserved vector current (Noether current).

Let us illustrate our rather formal procedure by looking at our simple Lagrangian (14.1):

$$\mathcal{L} = \overline{\psi}(x)(i\gamma \cdot \partial - m)\psi(x),$$  (14.28)

but now under an infinitesimal local phase transformation,

$$\psi(x) \rightarrow (1 + i\alpha(x))\psi(x),$$
$$\overline{\psi}(x) \rightarrow \overline{\psi}(x)(1 - i\alpha(x)), \alpha(x) \quad \text{inf initesimal}$$  (14.29)

so that we can generate the (Noether) vector current without the use of equations of motion.

Hence, let us consider the variation of (14.28) under the phase change (14.29):

$$\mathcal{L} \rightarrow \mathcal{L}[\alpha] = \overline{\psi}(x)\left(1 - i\alpha(x)\right)(i\gamma \cdot \partial - m)(1 + i\alpha(x))\psi(x) + \mathcal{O}(\alpha^2)$$
$$= \mathcal{L} - (\partial^\mu \alpha(x))\overline{\psi}\gamma_\mu \psi + \mathcal{O}(\alpha^2),$$

which yields

$$\frac{\delta \mathcal{L}[\alpha]}{\delta \alpha} = 0$$  (14.30)

and

$$\frac{\delta \mathcal{L}[\alpha]}{\delta (\partial^\mu \alpha)} = -\overline{\psi}\gamma_\mu \psi.$$  (14.31)

From (14.23) we obtain

$$J_\mu = \overline{\psi}\gamma_\mu \psi.$$  (14.32)

According to (14.30), the right-hand side of (14.24) vanishes, showing that $J_\mu$ is conserved:

$$\partial^\mu J_\mu = \partial^\mu \left( \overline{\psi} \gamma_\mu \psi \right) = 0. \tag{14.33}$$

At this stage, it is convenient to make contact with Schwinger's action principle applied to the phase change in the original fermionic Lagrangian:

$$\mathcal{L} = \overline{\psi}(x)(i\gamma \cdot \partial - m)\psi(x) = \mathcal{L}[0] \tag{14.34}$$

$$\psi \rightarrow \psi' = e^{i\alpha(x)}\psi,$$
$$\overline{\psi} \rightarrow \overline{\psi}' = \overline{\psi}e^{-i\alpha(x)}, \quad \alpha(x) \text{ finite!} \tag{14.35}$$

Under this finite phase change, we obtain

$$\mathcal{L}[0] \rightarrow \mathcal{L}[\alpha] = \overline{\psi}e^{-i\alpha}(i\gamma \cdot \partial - m)e^{i\alpha}\psi$$
$$= \overline{\psi}e^{-i\alpha}e^{i\alpha}(-\gamma \cdot \partial\alpha + i\gamma \cdot \partial - m)\psi$$
$$= \overline{\psi}(i\gamma \cdot \partial - m)\psi - \left( \overline{\psi}\gamma_\mu\psi \right)\partial_\mu\alpha$$
$$\mathcal{L}[\alpha] = \mathcal{L}[0] - J^\mu\partial_\mu\alpha. \tag{14.36}$$

This equation reveals the role of $\alpha(x)$ as an external source to which the current $J^\mu$ is coupled.

Now let us rewrite (14.36) in the form

$$\delta_\alpha\mathcal{L} = \mathcal{L}[\alpha] - \mathcal{L}[0] = -J^\mu\partial_\mu\alpha \tag{14.37}$$

and look at the response of $\alpha\alpha + \delta\alpha$, $\delta\alpha$ infinitesimal.

The result is

$$\delta_{\alpha+\delta\alpha}\mathcal{L} = -J^\mu\left(\partial_\mu\alpha + \partial_\mu(\delta\alpha)\right)$$
$$= \delta_\alpha\mathcal{L} - J^\mu\partial_\mu\delta\alpha$$

or

$$\delta\mathcal{L} = \delta_{\alpha+\delta\alpha}\mathcal{L} - \delta_\alpha\mathcal{L} = -J^\mu\partial_\mu\delta\alpha,$$

which we could have obtained from (14.36) by replacing $\alpha\delta\alpha$. Hence, if $\mathcal{L}$ changes according to

$$\mathcal{L} \rightarrow \mathcal{L} + \delta\mathcal{L},$$

then the action responds likewise:

$$W_{12} \rightarrow W_{12} + \delta W_{12}$$

where

$$\delta W_{12} = \int_{\sigma_2}^{\sigma_1} d^4x \,\delta\mathcal{L} = -\int_{\sigma_2}^{\sigma_1} d^4x\, J^\mu(\partial_\mu \delta\alpha)$$

$$= -\int_{\sigma_2}^{\sigma_1} d^4x\, \partial_\mu(J^\mu \delta\alpha) + \int_{\sigma_2}^{\sigma_1} d^4x\, \delta\alpha\,\partial_\mu J^\mu,$$

and $\sigma_{1,2}$ are space-like hypersurfaces.

If $\delta W_{12}$ is stationary with respect to an infinitesimal change in $\alpha$, then

$$\partial_\mu J^\mu = 0 \tag{14.38}$$

and $\delta W_{12}$ depends only on the endpoints

$$\delta W_{12} = -\int_{\sigma_2}^{\sigma_1} d\sigma_\mu J^\mu \delta\alpha = G[\sigma_1] - G[\sigma_2] \tag{14.39}$$

where

$$G[\sigma] = -\int^{\sigma} d\sigma_\mu J^\mu \delta\alpha \tag{14.40}$$

or, in flat space, $d\sigma_0 = d^3\vec{x}$

$$G[t] = -\int d^3\vec{x}\, J^0(\vec{x}, t)\delta\alpha(\vec{x}, t). \tag{14.41}$$

In particular, for constant $\delta\alpha$:

$$G[t] = -\delta\alpha \int d^3\vec{x}\, J^0(\vec{x}, t) = -\delta\alpha\, Q(t), \tag{14.42}$$

where $Q$ denotes the total charge with $\dot{Q}(t) = 0$.

If we substitute (14.41), with indices $i$ again, we can now rewrite (14.17) as

$$\frac{1}{i}[\psi_i, G] = \delta\psi_i = i\delta\alpha^l(x)T^l_{ij}\psi_j(x)$$

$$i\left[\psi_i(x), \int d^3\vec{x}' \, J^{0l}\left(\vec{x}', x^{0\prime}\right)\delta\alpha^l\left(\vec{x}', x^{0\prime}\right)\right] = i \int d^3\vec{x}' \delta^{(3)}\left(\vec{x} - \vec{x}'\right)\delta\alpha^l\left(\vec{x}', x^0\right)T_{ij}^l\psi_j\left(\vec{x}', x^0\right),$$

$$(14.43)$$

so that we obtain the important equal-time commutator

$$\left[J^{0l}(x), \psi_i(x')\right]_{x^{0\prime}=x^0} = -\delta^{(3)}\left(\vec{x} - \vec{x}'\right)T_{ij}^l\psi_j(x)  \qquad (14.44)$$

For constant $\delta\alpha^l$, (14.43) reduces to

$$\frac{1}{i}\left[Q^l(t), \psi_i(x)\right]\delta\alpha^l = i\delta\alpha^l T_{ij}^l\psi_j(x)$$

or

$$\left[Q^l(t), \psi_i(x)\right]_{x^0=t} = -T_{ij}^l\psi_j(x).  \qquad (14.45)$$

The infinitesimal version of (14.41) suggests that the following unitary transformation $U(t)$ generated by $G$ implies rotations in the internal space spanned by the $\psi_i's$:

$$U(t) = e^{iG_\alpha(t)}, U^\dagger = U^{-1},  \qquad (14.46)$$

where $\alpha(\vec{x}, t)$ is now finite! Then the transformed field operator is

$$(J_\alpha^0 \equiv J^{0l}\alpha^l)$$

$$\psi_i' = U\psi_i U^{-1} = e^{-i\int d^3\vec{x}' \, J^0\left(x'\right)\alpha\left(x'\right)}\psi_i\left(\vec{x}, x^0\right)e^{i\int d^3\vec{x}' \, J^0\left(x'\right)\alpha\left(x'\right)},  \qquad (14.47)$$

where all quantities are taken at the same time, $x^0 = x^{0\prime}$.
The evaluation of (14.47) yields:

$$\begin{aligned}
\psi_i'(x) &= \left(e^{i\alpha^l T^l}\right)_{ij}\psi_j(\vec{x}, t) \\
&= (\delta_{ij} + i\alpha^l T_{ij}^l + \ldots)\psi_j(\vec{x}, t) \\
&= \psi_i(\vec{x}, t) + i\alpha^l(\vec{x}, t)T_{ij}^l\psi_j(\vec{x}, t) + \mathcal{O}(\alpha^2),  \qquad (14.48)
\end{aligned}$$

which brings us back to our original Eq. (14.17).

This, then, concludes our proof that the unitary transformation (14.46) whose generator is $G_\alpha$ generates rotations in the internal (charge) space.

We can generalize our findings so far by presenting the following scheme for constructing a local gauge invariant theory. Let $\mathcal{L}(\Phi_k, \partial_\mu \Phi_k)$ be a globally gauge invariant Lagrangian, i.e.,

$$\mathcal{L}\left(\Phi', \partial_\mu \Phi'\right) = \mathcal{L}\left(\Phi, \partial_\mu \Phi\right)$$

with $\Phi' = \Omega\Phi$, $\Omega = $ const.

The transformations form a non-abelian group $G$ which is assumed to be semi-simple and compact, e.g., $G = SU(N)$, etc. The number of generators of $G$ is taken to be $n$, e.g., $n = N^2 - 1$ for $SU(N)$. Then we introduce a vector gauge field $A^i_\mu$ and define a covariant derivative:

$$D_\mu \Phi = \left(\partial_\mu - ig A^i_\mu T^i\right)\Phi, \tag{14.49}$$

with the second-rank field strength tensor:

$$F^i_{\mu\nu} = \partial_\mu A^i_\nu - \partial_\nu A^i_\mu + g f_{ijk} A^j_\mu A^k_\nu. \tag{14.50}$$

The local transformation properties are contained in $(A_\mu = T^i A^i_\mu)$

$$\Phi' = \Omega\Phi$$

$$A'_\mu = \Omega A_\mu \Omega^{-1} - \frac{i}{g}\left(\partial_\mu \Omega\right)\Omega^{-1} \tag{14.51}$$

$$F'_{\mu\nu} = \Omega F_{\mu\nu} \Omega^{-1}$$

with

$$\Omega = \exp\left\{-i\Theta^i T^i\right\} \tag{14.52}$$

The complete gauge invariant Lagrangian which describes the interaction between the gauge field $A^i_\mu$ and the fields $\Phi_k$ is then given by

$$\mathcal{L} = \mathcal{L}\left(\Phi_k, D_\mu \Phi_k\right) - \frac{1}{4} F^i_{\mu\nu} F^{i\mu\nu}. \tag{14.53}$$

If the group happens to be simple, there exists only one coupling constant. If, however, the group is a product of simple groups such as $SU(2) \times SU(2)$, where the respective set of generators closes under commutation with other sets, there exist independent coupling constants for each group. Here are some well-known examples:

(1) $G = U(1)$, electrodynamics

$\mathcal{L} = \overline{\psi}(i\gamma \cdot \partial - m)\psi$ is invariant under $\psi' = \Omega\psi = e^{ie\Lambda}\psi$, $\Lambda = $ const.

The number of generators is one, therefore, we have one gauge field, $A_\mu$. The covariant derivative is given by $e = g$)

$$D_\mu = \partial_\mu - ie A_\mu.$$

Since $f_{ijk} = 0$, we have $F_{\mu\nu} = \partial_\mu A_\nu - \partial_\nu A_\mu$.

Particles and gauge fields are then coupled according to the locally gauge invariant Lagrangian:

$$\mathcal{L} = \overline{\psi}(i\gamma \cdot D - m)\psi - \frac{1}{4}F_{\mu\nu}F^{\mu\nu}$$

(2) $G = SU(2)$, $\psi = \binom{p}{n}$

$\mathcal{L} = \overline{\psi}(i\gamma \cdots \partial - m)\psi$ is invariant under the gauge transformation of. the first kind:

$$\psi' = \Omega\psi, \quad \Omega \in SU(2), \quad \Omega = \text{const.}$$
$$\Omega = e^{\frac{i}{2}\vec{\tau}\cdot\vec{\Theta}}. \tag{14.54}$$

Since the number of generators is three, we need three gauge fields to gauge the isotopic spin.

Covariant derivative:

$$D_\mu \psi_k = (\partial_\mu \delta_{kl} - ig\frac{\tau^i_{kl}}{2}A^i_\mu)\psi_l \tag{14.55}$$

Field strength tensor:

$$F^i_{\mu\nu} = \partial_\mu A^i_\nu - \partial_\nu A^i_\mu + g\varepsilon_{ijk}A^j_\mu A^k_\nu. \tag{14.56}$$

Gauge invariant Lagrangian:

$$\mathcal{L} = \overline{\psi}(i\gamma \cdot D - m)\psi - \frac{1}{4}F^i_{\mu\nu}F^{i\mu\nu}$$
$$= \overline{\psi}(i\gamma \cdot \partial - m)\psi - \frac{1}{4}F^i_{\mu\nu}F^{i\mu\nu} + g\overline{\psi}_k\frac{\tau^i_{kl}}{2}A^i_\mu\gamma^\mu\psi_l. \tag{14.57}$$

$$G = SU(3), \text{ (2) chromodynamics } \psi = \begin{array}{|c|} \hline \psi\ red \\ \psi\ yellow \\ \psi\ blue \\ \hline \end{array}.$$

The Lagrangian of free quarks with three colors,

$$\mathcal{L} = \overline{\psi}(i\gamma \cdot \partial - m)\psi, \tag{14.58}$$

is invariant under

$$\psi \to \psi' = \Omega\psi, \Omega \in SU(3), \Omega = \text{const.}$$

$$\Omega = \exp\left\{\frac{i}{2}\sum_{i=1}^{8}\Theta^{i}\lambda^{i}\right\}, \quad T^{i} = \frac{\lambda^{i}}{2} \tag{14.59}$$

Covariant derivative:

$$D_{\mu}\psi_{k} = \partial_{\mu}\psi_{k} - ig\frac{\lambda^{i}_{kl}}{2}A^{i}_{\mu}\psi_{l}$$
$$\equiv (\partial_{\mu} - igA_{\mu})_{kl}\psi_{l} \tag{14.60}$$

To gauge the color group $SU(3)$, we need eight gauge vector fields (gluon fields, $A_{\mu} = A^{i}_{\mu}\frac{\lambda^{i}}{2}$).

The field strength tensor is

$$F^{i}_{\mu\nu} = \partial_{\mu}A^{i}_{\nu} - \partial_{\nu}A^{i}_{\mu} + gf_{ijk}A^{j}_{\mu}A^{k}_{\nu} \tag{14.61}$$

and the gauge invariant Lagrangian

$$\mathcal{L}^{QCD} = \overline{\psi}(i\gamma\cdot D - m)\psi - \frac{1}{4}F^{i}_{\mu\nu}F^{i\mu\nu}$$
$$\mathcal{L}^{QCD} = \overline{\psi}(i\gamma\cdot\partial - m)\psi - \frac{1}{4}F^{i}_{\mu\nu}F^{i\mu\nu} + g\overline{\psi}_{k}\frac{\lambda^{i}_{kl}}{2}A^{i}_{\mu}\gamma^{\mu}\psi_{l}. \tag{14.62}$$

The quark triplet forms the basis for the three-dimensional (fundamental) representation of $SU(3)$; the generators are the Gell-Mann matrices $T^{i} = \frac{\lambda^{i}}{2}$.

The field strength tensor $F^{i}_{\mu\nu}$, however, transforms according to the regular (adjoint) representation of $SU(3)$ :

$$(T^{a}_{adj})^{bc} = -if^{abc} \tag{14.63}$$

$a, b, c = 1, 2, \ldots n = (N^{2} - 1)$ for $SU(N) = 8$ for $SU(3)$.

(For $SU(2)$, it is the pion triplet that transforms according to the regular representation of the Lie algebra $[T^{i}, T^{j}] = i\varepsilon_{ijk}T_{k}, (T^{i}_{adj})^{kl} = -i\varepsilon^{ikl}$).

# Chapter 15
# Effective Actions (Lagrangians) in Quantum Field Theory

Most of the relativistic quantum field theories start out with a highly speculative object: the quantum vacuum. What it is and what is it good for, or what can we do without it? Has anybody measured it directly? Is it empty or is it filled with all kinds of (virtual) particles and antiparticles that live only a tiny fraction of time, too short to ever be detected?

It looks like we will never be able to measure it. It is simply an illusion and not a physical system. We have never been able to calculate the mass, the charge, etc. of any elementary particle within our existing quantum field theories. All these attributes which characterize a particle have to be taken from experiment. Therefore, we are using as a basis for all our calculations in quantum field theory a substratum that nobody knows anything about. It is like accepting the axioms of Euclid's geometry without knowing anything about space. It is certainly not something given a priori, as postulated by E. Kant.

Another point to mention is this: All of our local quantum field theory books use products of field operators to treat many-particle systems and their interactions. But all these processes need products of field operators at the same space–time point, which are highly singular; besides, to obtain finite results, one has to renormalize them, i.e., when changing from the field to the particle picture.

But there is a way out of our present limited understanding of fundamental field theories. However, for this we have to give up the idea of a local operator quantum field theory from which we can calculate all kinds of processes in high-energy physics using the mysterious vacuum. The answer to this problem is: We have to replace our so-called fundamental theories by *effective field theories*. This term sounds less demanding, and some of the more mathematically inclined theoretical physicists entertain the opinion that it is not even a theory. Our goal is to refute exactly this claim.

As our first example we will return to the Einstein-Hilbert theory of gravity which is based on Riemann's ideas (60 years earlier), which emphasize that we need a metric field in order to comprehend space. Our objects are c-number quantum fields (not

© The Author(s), under exclusive license to Springer Nature Switzerland AG 2021     111
W. Dittrich, *The Development of the Action Principle*,
SpringerBriefs in Physics,
https://doi.org/10.1007/978-3-030-69105-9_15

operator fields), Green's functions, etc. In analogy to Riemann, we start with the so-called "vacuum." We want to stick to this word, but, of course, with a completely different meaning, namely, an emptiness devoid of any particles, energy, etc. This nothingness is the background against which physical properties will take place. It helps to think of this vacuum as an extended manifold which obtains its structure (the physical space) as a result of perturbation by a source (apparatus) that is brought from the outside and is capable of producing all kinds of particles. In other words, only the coupling of the source to the vacuum can yield information about the response of the otherwise empty system through the generation of particles and any other process like, e.g., pair production.

Again, before the action of the source (accelerator), there is no particle or anything else in the vacuum. Only after the source has acted on the substratum vacuum will all the necessary attributes, like mass, charge, spin, etc. have been created by the transfer of energy, momentum and spin from the source to the vacuum.

Let us again emphasize: The fundamental description of a particle is its creation. For instance, a single particle can be created by a collision process in an accelerator. All the other particles function as a source for supplying the necessary energy, etc. to the ejected particle. We do not know the details of how the particle in question comes into existence; we only know that it is created through the effectiveness of a source $S(x)$ which is a numerical measure that describes where the collision act occurs, distributed in a certain region of space–time. The same degree of control in momentum space is written as $S(p)$, related to $S(x)$ by a Fourier transform. $S(p)$ is a representation of the fact that in some part of momentum space a particle is created.

Here, then, is our favorite philosophy which was baptized by Schwinger as *source theory*:

1.  Start with a state $|0_- >$ before anything happens.
2.  Thereafter bring in the source from the outside that supplies the necessary energy and all the other (symmetry-satisfying) quantum numbers for generating and selecting particles of physical interest for further investigations.
3.  Hence, it is the c-number source that stands at the beginning, which is thus coupled to the c-number field, which represents the field excitation (particle) and finally propagates with real measured attributes to the "sink" where it is detected in an absorption process. The state after everything is over is denoted by $< 0_+|$. No renormalization is necessary. The only condition is that free particle exchange has to be recovered when the interaction between particles travelling from source to sink is switched off. For higher order calculation, it is sometimes very hard to determine the necessary contact (or counter) terms to generate finite results.

Alternatively, we also make occasional use of Feynman's methods. In the source-field coupling of an interacting field theory, the difference between Schwinger's and Feynman's approaches is the following. In Schwinger's earlier functional field-theoretical formulation of QED, the external c-number sources, $\eta, \bar{\eta}, j$, coupled to the **operator** fields, $\psi, \overline{\psi}, A$, were set equal to zero in case the vacuum stays unchanged. Feynman, on the other hand, uses the c-number fields, $\psi, \overline{\psi}, A$, which have to be integrated over in the path-integral representation, since they are not observed! (Remember that in classical mechanics, the fast dynamical variables are integrated out and are treated as a background relative to which the slow degrees of freedom develop.)

In either case, the vacuum-persistence amplitude in the one-loop approximation is given by

$$< 0_+ |0_- >^{A^{ext}} = e^{i W^{(1)}[A^{ext}]} = exp\left\{-Tr \ln\left(\frac{G_+[A^{ext}]}{G_+[0]}\right)\right\} = e^{i \int d^4x \mathcal{L}^{(1)}(x)},$$

where the effective Lagrangian $\mathcal{L}^{(1)}(x)$ is the formal expression for the effect which an arbitrary number of "external photon lines" (classical electromagnetic fields) can have on a simple fermion loop.

To become more familiar with the strategies just described, let us study the conventional procedure, where a quantized matter field, a field operator $\Phi(x)$ is coupled to an external classical background gravitational field $g_{\mu\nu}(x)$. Then the action is given by

$$W_\Phi = \frac{1}{2} \int d^4x \sqrt{-g(x)} \cdot \{-g^{\mu\nu}(x)\partial_\mu\Phi\partial_\nu\Phi - m^2\Phi^2\}, \qquad (15.1)$$

where the metric tensor field $g^{\mu\nu}$ is a prescribed function of $x$. (The most general Lagrangian for $\Phi$ would also contain a term $\sim R\Phi^2$ with $R$ being the scalar curvature.)

Again, (15.1) defines a semiclassical theory, where $g^{\mu\nu}(x)$ is treated classically, whereas $\Phi(x)$ is treated quantum mechanically as a field operator.

Then the vacuum expectation of the matter field energy–momentum tensor,

$$T_{\mu\nu} = \partial_\mu\Phi\partial_\nu\Phi - \frac{1}{2}g_{\mu\nu}g^{\sigma\varrho}(x)\partial_\sigma\Phi\partial_\varrho\Phi$$

acts as a source on the right-hand side of Einstein's equations:

$$R_{\mu\nu} = -\frac{1}{2}g_{\mu\nu}R + \Lambda g_{\mu\nu} = -8\pi G\langle 0|T_{\mu\nu}|0\rangle. \qquad (15.2)$$

We know already that Eq. (15.2), with the right-hand side set equal to zero, is obtained as the variation of the usual Einstein–Hilbert action.

$$W^{(0)}[g] = \frac{1}{16\pi G} \int d^4x \sqrt{-g}(R - 2\Lambda), \qquad (15.3)$$

$$\frac{2}{\sqrt{-g}} \frac{\delta W^{(0)}[g]}{\delta g^{\mu\nu}(x)} = 0 \qquad (15.4)$$

If we now define the effective action $W^{(1)}$ by

$$\frac{2}{\sqrt{-g}} \frac{\delta W^{(1)}[g]}{\delta g^{\mu\nu}(x)} = \langle 0|T^{\mu\nu}(x)|0\rangle, \qquad (15.5)$$

our generalized Eq. (15.2) are derived by

$$\frac{2}{\sqrt{-g}} \frac{\delta}{\delta g^{\mu\nu}(x)} \left(W^{(0)} + W^{(1)}\right) = 0 \qquad (15.6)$$

However, in contrast to Maxwell's equations, Einstein's equations are highly non-linear, already at the purely classical level. But the strategy is the same in both cases: because one does not want to treat the microscopic degrees of freedom of the bosonic quantum fields $\Phi(x)$ or the fermionic fields $\Psi(x)$ explicitly, one derives effective equations of motion simulating their presence for the macroscopic, classical field $A_\mu(x)$ or $g_{\mu\nu}(x)$.

For the Eq. (15.2) to be consistent, $<0|T^{\mu\nu}(x)|0>$ must be covariantly conserved, because the left-hand side of (15.2) is.$<0|T^{\mu\nu}(x)|0>$ is indeed covariantly conserved if $W^{(1)}$ is invariant under general coordinate transformations, just as it is necessary for $W^{(1)}$ of electrodynamics to be gauge invariant for the induced vacuum current to be conserved. This is one example of the correspondence between gauge invariance (of the second kind) in electrodynamics and general covariance in gravitation theory. Both are dynamical symmetries, which tell us how particles couple to the external fields.

Let us compare these results in more detail with quantum electrodynamics, where quantized electron–positron fields (Dirac) are coupled to an external, i.e., unquantized electromagnetic field. In the source-free regions of space–time ($j_\mu = 0$) this external field obeys in absence of matter fields the classical Maxwell equations $\partial_\mu F^{\mu\nu} = 0$, which can be derived from the variation principle,

$$\frac{\delta W^{(0)}}{\delta A^\mu(x)} = \frac{\delta}{\delta A^\mu(x)} \int d^4x' \mathcal{L}^{(0)}(x') = 0,$$

$$\mathcal{L}^{(0)} = -\frac{1}{4} F_{\mu\nu} F^{\mu\nu}, \quad F_{\mu\nu} = \partial_\mu A_\nu - \partial_\nu A_\mu.$$

Our aim is to find an effective action,

$$W_{eff}[A] = W^{(0)}[A] + W^{(1)}[A], \qquad (15.7)$$

where $W^{(1)}$ describes the non-linear effects induced by the quantized fermion fields. The new equations of motion are then given by

$$\frac{\delta W_{eff}[A]}{\delta A^\mu(x)} = 0. \tag{15.8}$$

Alternatively, we want to study quantized fermions in a classical background electromagnetic field, which is now considered from the path integral point of view. For the moment, it suffices to look at an arbitrary field theory with fields $\{\Phi\}$ and Lagrangian $\mathcal{L}(\{\Phi\})$. Transition amplitudes then can be expressed as functional integrals of the general form

$$Z = \int [d\{\Phi\}] exp(iS[\{\Phi\}]) \tag{15.9}$$

with the action $S[\{\Phi\}] = \int d^4x \mathcal{L}(\{\Phi\})$.

Now assume that the set $\{\Phi\}$ can be divided in two subsets $\{\Phi^L\}$ and $\{\Phi^H\}$, where $\{\Phi^L\}$ are "light" field components whose dynamics we directly observe (the photon field, or the classical $A_\mu(x)$, in our case), while $\{\Phi^H\}$ are "heavy" fields (the electron field in our case) which influence the dynamics of the light fields, but are not directly observable. Since the $\{\Phi^H\}$ are hidden from view, it is convenient to write (15.9) in the form

$$Z = \int [d\{\Phi^L\}d\{\Phi^H\}] \exp(iS[\{\Phi^L\}, \{\Phi^H\}])$$
$$= \int [d\{\Phi^L\}] exp(iW_{eff}[\{\Phi^L\}]), \tag{15.10}$$

where the effective action $W_{eff}$ for the light fields is defined by

$$exp(iW_{eff}[\{\Phi^L\}]) = \int [d\{\Phi^H\}] \exp(iS[\{\Phi^L\}, \{\Phi^H\}]). \tag{15.11}$$

Clearly, the effective action, if exactly known, gives a complete description of the dynamics of $\{\Phi^L\}$ without any reference to the heavy fields.

Now let us consider fermions in an external electromagnetic field. Then we have

$$W_{eff} = W^{(0)} + W^{(1)},$$

$$W^{(0)} = \int d^4x \left\{ -\frac{1}{4} F_{\mu\nu} F^{\mu\nu} \right\},$$

and for $W^{(1)}$:

$$\exp(iW^{(1)}[A]) = \int [d\psi d\overline{\psi}] exp\left[ (-i\int d^4x\, \overline{\psi} \left\{ \gamma^\mu \left( \frac{1}{i}\partial_\mu - e A_\mu \right) - m \right\} \psi \right]. \tag{15.12}$$

According to the general rules for the path integral quantization of Fermi fields, $\psi$ and $\overline{\psi}$ are anti-commuting classical fields forming a Grassmann algebra; hence, a

Gauss-type integral like (15.12) can be evaluated to be

$$\exp\bigl(i\,W^{(1)}[A]\bigr) = det\left[\gamma^{\mu}\left(\frac{1}{i}\partial_{\mu} - eA_{\mu}\right) + m\right] = det\bigl(G[A]^{-1}\bigr). \qquad (15.13)$$

This gives

$$W^{(1)}[A] = -i\ln det\bigl(G[A]^{-1}\bigr) = +i\ln det\,G[A] = +i\,Tr\ln G[A], \qquad (15.14)$$

where we used the (formal) identity *det (exp G) = exp (Tr G)*. Because action functionals are defined only up to a constant, we may exploit this freedom to replace (15.14) by

$$W^{(1)}[A] = i\,Tr\ln G[A] - i\,Tr\ln G[0],$$

$$= i\,Tr\ln\left(\frac{G[A]}{G[0]}\right). \qquad (15.15)$$

# Chapter 16
# Modified Photon Propagation Function, Source Theory

This chapter is almost exclusively concerned with quantum electrodynamics (QED). The emphasis on this well-explored topic serves conveniently to elaborate the viewpoints and techniques of the so-called source theory. We intend to describe a self-contained development of the process whereby an extended photon source permits emission or absorption of a neutral pair of charged particles. It is the "source" that produces a time-like virtual photon ($k^2 = -M^2$) which, in turn, transfers the necessary energy to a real on-shell particle-antiparticle pair.

There are no Feynman graphs involved. Our fundamental graph (Fig. 16.1) is a "causal graph." The exchanged particles travel from the source to the sink with their real, observed masses. There is no renormalization procedure necessary, neither for the mass nor for the charge. Here, then, is the corresponding analysis that takes us from the free traveling photon ($k^2 = 0$) to the modified effective photon propagator which experiences from the source an excess of energy ($k^2 = -M^2$) so that after an extremely short time, it can produce an electron–positron pair. Everything happens between the "vacua" $<0_+|$ and $|0_->$. These are not the vacua that are boiling with particle-antiparticle pairs, etc. They are absolutely empty until an external source delivers or takes the necessary attributes of energy, momentum, spin, etc. to or from the particles to be produced or annihilated.

Here, then, is the detailed calculation leading to the modified photon propagator which enables us at the same time to find the altered Coulomb potential or the Uehling potential of QED. There are no priority claims! But part of the way the calculation was done is new.

Now let us begin with the QED action,

$$W = \int (dx) \left\{ J^\mu A_\mu + \eta \gamma^0 \psi - \frac{1}{4} F^{\mu\nu} F_{\mu\nu} - \frac{1}{2} \psi \gamma^0 \left[ \gamma^\mu \left( -\frac{1}{i} \partial_\mu - eq A_\mu \right) + m \right] \psi \right\} \quad (16.1)$$

The $\psi$'s are real, in Majorana presentation; $q = \begin{pmatrix} 0 & -i \\ i & 0 \end{pmatrix}$ charge metric matrix.

© The Author(s), under exclusive license to Springer Nature Switzerland AG 2021
W. Dittrich, *The Development of the Action Principle*,
SpringerBriefs in Physics,
https://doi.org/10.1007/978-3-030-69105-9_16

We also recall the vacuum amplitude (V.A.)

$$\langle 0_+|0_- \rangle^{\eta,J} = e^{iW} = e^{\frac{i}{2}\int (dx)(dx')\eta(x)\gamma^0 G_+(x-x')\eta(x')}e^{\frac{i}{2}\int (dx)(dx')J^\mu(x)D_+(x-x')J_\mu(x')}$$

$$= e^{\frac{i}{2}\int \eta\gamma^0\psi}e^{\frac{i}{2}\int J^\mu A_\mu}$$

$$\equiv e^{i2W-iW} = e^{i\int \eta\gamma^0\psi}e^{i\int J_\mu A^\mu}e^{-i\int \mathcal{L}(\psi,A)},$$

where

$$\mathcal{L} = -\frac{1}{2}\psi\gamma^0\left[\gamma^\mu\left(-\frac{1}{i}\partial_\mu - eqA_\mu\right) + m\right]\psi - \frac{1}{4}(F_{\mu\nu})^2. \qquad (16.2)$$

From the action principle, we obtain the field equation

$$\left[\gamma\left(\frac{1}{i}\partial - eqA\right) + m\right]\psi^A = \eta^A$$

or the Green's function equation:

$$\left[\gamma\left(\frac{1}{i}\partial - eqA\right) + m\right]G_+\left(x, x'|A\right) = \delta\left(x - x'\right). \qquad (16.3)$$

The V.A. that refers to the primitive interaction is

$$V.A. = i\int (dx)A^\mu \frac{1}{2}\psi\gamma_0 eq\gamma_\mu\psi \equiv i\int (dx)A_\mu j^\mu,$$

$$j^\mu(x) = \frac{1}{2}\psi(x)\gamma_0 eq\gamma_\mu\psi(x). \qquad (16.4)$$

The expression for the source in emission is given by

$$i\eta(x)\eta\left(x'\right)|_{eff.em} = \delta\left(x - x'\right)eq\gamma^\mu\gamma^0 A_\mu(x). \qquad (16.5)$$

$\eta$ carries spin and charge indices, i.e., is a $4 \times 2 = 8$-component object. Both sides of (16.5) are anti-symmetrical.

In momentum space:

$$i\eta(p)\eta\left(p'\right)|_{eff.em} = eq\gamma^\mu\gamma^0 A_\mu(k)k = p + p'. \qquad (16.6)$$

In the following we want to discuss the modification of the photon propagation function, emission (and absorption) of two non-interacting particles which propagate freely between the effective sources in Fig. 16.1, as described by the quadratic term of

**Fig. 16.1** Causal graph of
modified photon propagator

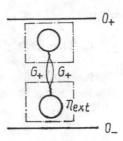

$$\langle 0_+ | 0_- \rangle = \exp\left\{ i \int \eta_1 \gamma^0 G_+ \eta_2 \right\} = \exp\left\{ i \eta_1(x) \gamma^0 i \int d\omega_p e^{ip(x-x')} (m - \gamma p) \eta_2(x') \right\}$$

$$= \exp\left\{ - \int d\omega_p \eta_1(-p) \gamma^0 (m - \gamma p) \eta_2(p) \right\}. \tag{16.7}$$

$$d\omega_p = \frac{d^3 \vec{p}}{(2\pi)^3} \frac{1}{2p_0}$$

Exchange of two particles between sources is then given by picking the quadratic term only:

$$< 0_+ | 0_- > = \frac{1}{2} \int d\omega_p d\omega_{p'} \eta_1(-p) \gamma^0 (m - \gamma p) \eta_2(p) \eta_2\left(p'\right) \gamma^0 \left(m + \gamma p'\right) \eta_1\left(-p'\right). \tag{16.8}$$

Using

$$\eta_1(-p)_a M_{ab} \eta_1\left(-p'\right)_b = -M_{ab} \eta_1\left(-p'\right)_b \eta_1(-p)_a = tr\left[ M \eta_1\left(-p'\right) \eta_1(-p) \right],$$

where $a, b$ are indices for the 8-dimensional component $\eta$, the trace thus being an 8-dimensional one. This yields

$$< 0_+ | 0_- > = \frac{1}{2} \int d\omega_p d\omega_{p'} tr\left[ (m - \gamma p) \eta_2(p) \eta_2\left(p'\right) \gamma^0 \left(-m - \gamma p'\right) \eta_1\left(-p'\right) \gamma^0 \right] \tag{16.9}$$

After inserting the effective sources (16.6), we obtain

$$< 0_+ | 0_- > = -\frac{1}{2} \int d\omega_p d\omega_{p'} tr\left[ (m - \gamma p) eq \gamma A_2(k) \left(-m - \gamma p'\right) eq \gamma A_1(-k) \right]. \tag{16.10}$$

Substituting the unit factor $1 = (2\pi)^3 \int dM^2 d\omega_k \delta\left(k - p - p'\right)$, we find

$$< 0_+ | 0_- > = -e^2 \int dM^2 d\omega_k A_1^\mu(-k) I_{\mu\nu}(k) A_2^\mu(k), \tag{16.11}$$

where

$$I_{\mu\nu}(k) = (2\pi)^3 \int d\omega_p d\omega_{p'} \delta(k - p - p') tr_4 [\gamma_\mu(m - \gamma p)\gamma_\nu(-m - \gamma p')],$$

which can be shown to be symmetrical in $\mu$ and $\nu$. $I_{\mu\nu}(k)$ carries a gauge invariant structure, as can be demonstrated by making a gauge transformation:

$$A_\mu(k) \to A_\mu(k) - i k_\mu \lambda(k) \tag{16.12}$$

and showing that $k^\mu I_{\mu\nu}(k) = 0$.

The symmetrical tensor constructed from the vector $k^\mu$ is

$$I_{\mu\nu}(k) = \left(g_{\mu\nu} - \frac{k_\mu k_\nu}{k^2}\right) I(M^2). \tag{16.13}$$

To find $I(M^2)$, consider the trace of $I_{\mu\nu}(k)$:

$$I_\mu^\mu = 3I(M^2) = \int d\omega_p d\omega_{p'} (2\pi)^3 \delta(p + p' - k) tr_4 [\gamma^\mu(m - \gamma p)\gamma_\mu(-m - \gamma p')]. \tag{16.14}$$

Using

$$\gamma^\mu \gamma_\mu = -4, \gamma^\mu \gamma \cdot p \gamma_\mu = 2\gamma p \quad \tfrac{1}{4} tr[\gamma_\mu \gamma_\nu] = -g_{\mu\nu}, tr\gamma_\mu = 0$$
$$k = p + p', k^2 = -M^2 = -2m^2 + 2pp',$$
$$(2\pi)^3 \int d\omega_p d\omega_{p'} \delta(p + p' - k) = \frac{1}{(4\pi)^2}\left(1 - \frac{4m^2}{M^2}\right)^{\frac{1}{2}}, \tag{16.15}$$

we obtain.

$$I(M^2) = \frac{4}{3}(M^2 + 2m^2)\frac{1}{(4\pi)^2}\left(1 - \frac{4m^2}{M^2}\right)^{\frac{1}{2}}. \tag{16.16}$$

Hence, the V.A. (vacuum amplitude) can be rewritten in the form

$$< 0_+ | 0_- > = -e^2 \int dM^2 d\omega_k A_1^\mu(-k)\left(g_{\mu\nu} - \frac{k_\mu k_\nu}{M^2}\right)\frac{4}{3}\frac{1}{(4\pi)^2}(M^2 + 2m^2)$$
$$\left(1 - \frac{4m^2}{M^2}\right)^{\frac{1}{2}} A_2^\nu(k) \tag{16.17}$$

Since this expression is gauge invariant, we have $k_\mu J_{1,2}^\mu = 0$ after having inserted

$$A_2^{\nu}(k) = -\frac{1}{M^2} J_2^{\nu}(k), \ A_1^{\mu}(-k) = -\frac{1}{M^2} J_1^{\mu}(-k). \tag{16.18}$$

The V.A. so far is then given by

$$< 0_+ |0_- >= i\frac{\alpha}{3\pi} \int \frac{dM^2}{M^2} \left(1 + \frac{2m^2}{M^2}\right) \sqrt{1 - \frac{4m^2}{M^2}} \, id\omega_k J_1^{\mu}(-k) J_2^{\nu}(k)$$

$$\int J_1^{\mu}(x) e^{ikx} (dx) \qquad \int e^{-ikx'} J_{2\mu}(x')(dx').$$

(16.19)

Substituting

$$i \int d\omega_k \exp\left(ik\left(x - x'\right)\right) = \Delta_+\left(x - x'; M^2\right), \tag{16.20}$$

we obtain at last

$$< 0_+ |0_- >= i\frac{\alpha}{3\pi} \int \frac{dM^2}{M^2} \left(1 + \frac{2m^2}{M^2}\right) \sqrt{1 - \frac{4m^2}{M^2}} \int (dx)(dx') J_1^{\mu}(x) \Delta_+$$

$$\left(x - x'; M^2\right) J_2^{\mu}\left(x'\right), \tag{16.21}$$

which modifies the free photon propagation function according to.

$$\overline{D}_+(k) = \frac{1}{k^2 - i\varepsilon} + \int dM^2 \frac{a\left(M^2\right)}{k^2 + M^2 - i\varepsilon}, \ a\left(M^2\right) = \frac{\alpha}{3\pi} \frac{1}{M^2} \left(1 + \frac{2m^2}{M^2}\right) \sqrt{1 - \frac{4m^2}{M^2}}. \tag{16.22}$$

The spectral weight function $a\left(M^2\right)$ is real and non-negative. For $M \gg 2m$, the integral behaves like

$$\int \frac{dM^2}{M^2} \frac{1}{k^2 + M^2}$$

and there is no question about the existence from threshold up to infinity. In configuration space, we obtain

$$\overline{D}_+(x - x') = D_+(x - x') + \frac{\alpha}{3\pi} \int_{(2m)^2}^{\infty} \frac{dM^2}{M^2} \left(1 + \frac{2m^2}{M^2}\right) \left(1 - \frac{4m^2}{M}\right)^{\frac{1}{2}} \Delta_+\left(x - x'; M^2\right)$$

Let us return to the V.A. (16.11) with $I_{\mu\nu}(k)$ as given in (16.13) and (16.16):

$$< 0_+|0_- > = -e^2 \int dM^2 I(M^2) d\omega_k A_1^\mu(-k) \left( g_{\mu\nu} + \frac{k_\mu k_\nu}{M^2} \right) A_2^\nu(k). \qquad (16.23)$$

The inherent gauge invariance can be made explicit be introducing the field strength tensor,

$$F_{\mu\nu}(k) = ik_\mu A_\nu(k) - ik_\nu A_\mu(k),$$

where

$$-\frac{1}{2} F_1^{\mu\nu}(-k) F_{2\mu\nu}(k) = M^2 A_1^\mu(k) \left( g_{\mu\nu} + \frac{k_\mu k_\nu}{M^2} \right) A_2^\nu(k), \qquad (16.24)$$

so that we obtain for (16.23)

$$= ie^2 \int \frac{dM^2}{M^2} I(M^2) \left( -\frac{1}{2} \right) F_1^{\mu\nu}(-k) i d\omega_k F_{2\mu\nu}(k). \qquad (16.25)$$

After space–time extrapolation, we obtain the action expression

$$\int dM^2 M^2 a(M^2)(-\frac{1}{4}) \int (dx)(dx') F^{\mu\nu}(x) \left[ \Delta_+(x - x'; M^2) + c.t. \right] F_{\mu\nu}(x'), \qquad (16.26)$$

with.

$$M^2 a(M^2) = \frac{4\pi\alpha}{M^2} I(M^2) = \frac{\alpha}{3\pi} \left( 1 + \frac{2m^2}{M^2} \right) \sqrt{1 - \frac{4m^2}{M^2}}. \qquad (16.27)$$

The contact term (c.t.) has to be chosen so as to recover the free photon propagation function ($k^2 = 0$) for the coupling $a(M^2) = 0$, i.e.,

$[\Delta_+(x - x'; M^2) + c.t.] = 0$ for $a(M^2) = 0$ or, in momentum space:

$$\left[ \frac{1}{k^2 + M^2 - i\varepsilon} + c.t. \right] = 0.$$

The minimum choice for keeping the correct pole structure ($k^2 = 0$) of the freely propagating photon is

$$\left[ \Delta_+\left(x - x'; M^2\right) \right] - \frac{1}{M^2} \delta\left(x - x'\right) = \frac{1}{M^2} \partial^2 \Delta_+(x - x'; M^2) \qquad (16.28)$$

or, in momentum space:

$$\frac{1}{k^2 + M^2 - i\varepsilon} - \frac{1}{M^2} = -\frac{k^2}{M^2}\frac{1}{k^2 + M^2 i\varepsilon}. \tag{16.29}$$

The consequence is a modified field equation with a modified vector potential:

$$\overline{A}^\mu(k) = \overline{D}_+(k)J^\mu(k), \tag{16.30}$$

where

$$\overline{D}_+(k) = \frac{1}{k^2 - i\varepsilon}\frac{1}{1 - k^2 \int dM^2 a(M^2)\frac{1}{k^2 + M^2 - i\varepsilon}} \tag{16.31}$$

$$a(M^2)small \cong \frac{1}{k^2 - i\varepsilon} + \int dM^2 \frac{a(M^2)}{k^2 + M^2 - i\varepsilon} \tag{16.32}$$

In x-space, a more complete action for the electromagnetic field is then given by

$$W = \int(dx)\left[J^\mu(x)A_\mu(x) - \frac{1}{4}F^{\mu\nu}(x)F_{\mu\nu}(x)\right]$$

$$- \int dM^2 a(M^2)(-\frac{1}{4})\int(dx)(dx')\partial^\lambda F^{\mu\nu}(x)\Delta_+(x - x'; M^2)\partial'_\lambda F_{\mu\nu}(x') \tag{16.33}$$

Here we lost the original free local interaction necessary to define a Lagrange function. But for slowly varying fields over the interval $\frac{1}{M} < \frac{1}{2m}$, we can simplify (16.33) by letting $x' \cong x$, so that

$$\int(dx')\Delta_+(x - x'; M^2 = \Delta_+(k^2 = 0; M^2) = \frac{1}{M^2}$$

and together with

$$\int_{(2m)^2}^\infty \frac{dM^2}{M^2}a(M^2) = \frac{\alpha}{(3\pi)}\int_{(2m)^2}^\infty \frac{dM^2}{(M^2)^2}(1 + \frac{2m^2}{M^2})\left(1 - \frac{4m^2}{M^2}\right)^{\frac{1}{2}}$$

$$v = \left(1 - \frac{4m^2}{M^2}\right)^{\frac{1}{2}} := \frac{\alpha}{\pi}\frac{1}{(2m)^2}\int_0^1 dv v^2\left(1 - \frac{1}{3}v^2\right) = \frac{\alpha}{15\pi}\frac{1}{m^2}, \tag{16.34}$$

we can replace the last term of (16.33) by

$$-\frac{\alpha}{15\pi}\frac{1}{m^2}\int(dx)(-\frac{1}{4})\partial^\lambda F^{\mu\nu}(x)\partial_\lambda F_{\mu\nu}(x). \tag{16.35}$$

In this local limit, we find again a Lagrange function:

$$\mathcal{L} = -\frac{1}{4}\left[ F^{\mu\nu}F_{\mu\nu} - \frac{\alpha}{15\pi}\frac{1}{m^2}\partial^\lambda F^{\mu\nu}\partial_\lambda F_{\mu\nu} \right].$$

As a consequence, we find a modified Maxwell field equation:

$$\left(1 + \frac{\alpha}{15\pi}\frac{1}{m^2}\partial^2\right)\partial_\nu F^{\mu\nu}(x) = J^\mu(x), \tag{16.36}$$

or the approximate solution ($\partial^2 \ll m^2$)

$$A_\mu(x) = \left(1 - \frac{\alpha}{15\pi}\frac{1}{m^2}\partial^2\right)\int (dx')D_+(x - x')J_\mu(x') \tag{16.37}$$

$$= \int (dx')D_+(x - x')J_\mu(x') + \frac{\alpha}{15\pi}\frac{1}{m^2}J_\mu(x) \tag{16.38}$$

up to a gauge term.

The explicit expression for $W$ is now given by

$$W = \frac{1}{2}\int(dx)J^\mu(x)A_\mu(x)$$
$$= \frac{1}{2}\int(dx)(dx')J^\mu(x)D_+(x - x')J_\mu(x') + \frac{\alpha}{15\pi}\frac{1}{m^2}\frac{1}{2}\int(dx)J^\mu(x)J_\mu(x),$$

which implies the modified interaction energy of two static charge-current distributions

$$E_{\text{int.}} = -\int(d\vec{x})(d\vec{x}')J_a^\mu(\vec{x})\mathcal{D}(\vec{x} - \vec{x}')J_{b\mu}(\vec{x}') - \frac{\alpha}{15\pi}\frac{1}{m^2}\int(d\vec{x})J_a^\mu(\vec{x})J_{b\mu}(\vec{x}) \tag{16.39}$$

Let us return for a moment to Eq. (16.31) and its approximation (16.32), which we write as

$$A_\mu^{eff}(k) = A_\mu^{ext}(k)\left(1 + k^2\frac{\alpha}{3\pi}\int_{(2m)^2}^\infty \frac{dM^2}{(M^2)}\left(1 + \frac{2m^2}{M^2}\right)\left(1 - \frac{4m^2}{M^2}\right)^{\frac{1}{2}}\frac{1}{k^2 + M^2 - i\varepsilon}\right)$$

$$v = \left(1 - \frac{4m^2}{M^2}\right)^{\frac{1}{2}} := A_\mu^{ext}(k)\left(1 + k^2\frac{\alpha}{4\pi}\int_0^1 dv \frac{v^2\left(1 - \frac{v^2}{3}\right)}{\left[m^2 + \frac{k^2}{4}\left(1 - v^2\right)\right]}\right)$$

For the non-relativistic case $\left(\frac{k^2}{m^2}\right) \ll 1$, the $v-$ integral gives $\int_0^1 dv\left(v^2 - \frac{v^4}{3}\right) = \frac{4}{15}$,

so that we find

$$A_\mu^{eff}(k) = [1 + \frac{\alpha}{15\pi m^2}k^2]A_\mu^{ext}(k)$$

or, in configuration space,

$$A_\mu^{eff}(x) = A_\mu^{ext}(x) - \frac{\alpha}{15\pi m^2}\square A_\mu^{ext}.$$

Hence, for any external electromagnetic field, there is always—because of pair production giving vacuum polarization—an effective electromagnetic field differing from the external field by the induced field.

For the special case of the Coulomb field of an H-like atom, we have

$$A_\mu^{ext}(x) = \left(A_0, \vec{A}\right) = \left(V^{ext}(x), \vec{0}\right), V^{ext}(r) = +(\frac{Ze}{4\pi r}),$$

and the effective field then given by

$$V^{eff}(r) = V^{ext}(r) + V^{ind}(r)$$

$$= (1 - \frac{\alpha}{15\pi m^2}\vec{\nabla}^2)\left(\frac{Ze}{4\pi r}\right)$$

and with the Green's function equation $\vec{\nabla}^2(\frac{1}{r}) = -4\pi\delta(\vec{r})$,

$$V^{eff}(r) = \frac{Ze}{4\pi r} + \frac{4\alpha}{15m^2}\frac{Ze}{4\pi}\delta(\vec{r}) = \frac{\alpha}{15m^2}\frac{Ze}{\pi}\delta(\vec{r}),$$

representing an augmentation of the Coulomb potential at the origin.

The contribution to the energy of an electron, charge $-e$, moving in such an extra induced field, is non-zero only for $s$ states (because of $\delta(\vec{r})$).

If we denote by $|a\rangle$ the state $|n,l\rangle$, we get a level shift due to vacuum polarization:

$$\Delta E_{nl} = -\langle a|eV^{ind}|a\rangle$$

$$= -\frac{\alpha Ze^2}{15\pi m^2}\int \psi_{nl}^*(\vec{r})\delta(\vec{r})\psi_{nl}(\vec{r})d^3r = -\frac{\alpha Ze^2}{15\pi m^2}|\psi_{nl}(0)|^2,$$

where $|\psi_{nl}(0)|^2 = \frac{1}{\pi}\left(\frac{Z\alpha}{n}\right)^3\left(\frac{mc}{\hbar}\right)^3\delta_{l0}$, i.e., only $s$ states.

$$\hbar = 1 = c: \quad = \frac{1}{\pi}\left(\frac{Z\alpha m}{n}\right)^3\delta_{l0}.$$

Hence, the energy levels of $s$ states, for principal quantum number $n$, are depressed by the amount of the vacuum polarization

$$\Delta E_{ns} = -\frac{\alpha Z e^2}{15\pi m^2}\frac{1}{\pi}\left(\frac{Z\alpha}{n}\right)^3 m^3, \quad e^2 = 4\pi\alpha$$

$$= -\frac{4}{15\pi}\frac{\alpha(Z\alpha)^4}{n^3}m$$

and for the s state ($n = 2, l = 0$) with $Z = 1$, we get a frequency shift:

$$\boxed{v = \frac{\Delta E_{\text{v.pol}}}{h} = -27mc/sec}$$

Therefore, the $2S_{\frac{1}{2}}$ and $2P_{\frac{1}{2}}$ levels, degenerate by the Dirac theory, should be shifted relative to each other with the S level lowered. However, the observed shift is approximately $1000\,mc/sec$ in the opposite direction (Lamb shift, etc.). Since agreement between theory and experiment is about $\frac{1}{2}\,mc/sec$, (i.e., $\sim\alpha(Z\alpha)^2$), we have the experimental verification of the vacuum polarization effect.

The amount of level shift from the vacuum polarization by giving a comparison to the energy values due to.

(1)   non-relativistic Bohr level $= \frac{1}{2}(Z\alpha)^2 mc^2$
(2)   fine structure (spin–orbit interaction) $\sim(Z\alpha)^4 mc^2$
(3)   s state Lamb shift $\sim\alpha(Z\alpha)^4 mc^2$.

Finally, we want to use the action principle to investigate the modification of Coulomb's Law for long distances between charged particles. (See [1, 2, 3].) First we will ask for the effective Lagrangian for weak but otherwise arbitrary fields. In the weak-field limit, $e^2 F_{\mu\nu} F^{\mu\nu}/m^4$ becomes small due to the smallness of $\alpha$. This leads to an $\alpha$-order expression ($wf$ = weak field) to the one-loop quantum correction to the classical electromagnetic field contribution:

$$W_{wf}^{(1)}[A] = \int \mathcal{L}_{wf}^{(1)}(dx) = \frac{1}{2}\int A^\mu(x)\Pi_{\mu\nu}(x, y)A^\nu(y)(dx)(dy) + \alpha^2 \quad (16.40)$$

From our former calculation, we can take over (with small changes of notation) the vacuum polarization tensor,

$$\Pi_{\mu\nu}(k) = \left(g_{\mu\nu}k^2 - k_\mu k_\nu\right)\Pi\left(k^2\right)$$

$$\Pi\left(k^2\right) = \frac{\alpha}{3\pi}k^2\int\limits_{(2m)^2}^{\infty}\frac{dt}{t}\left(1 + \frac{2m^2}{t}\right)\left(1 - \frac{4m^2}{t}\right)^{\frac{1}{2}}\frac{1}{t - k^2 - i\varepsilon}, \quad (16.41)$$

or, in space representation: $\Box e^{ikx} = -k^2 e^{ikx}, \Box = \frac{\partial}{\partial t^2} - \Delta,$

$$\Pi_{\mu\nu}(\Box) = (g_{\mu\nu}\Box - \partial_\mu\partial_\nu)\Pi(\Box),$$

$$\Pi(\Box) = -\frac{\alpha}{3\pi}\Box \int_{(2m)^2}^\infty \frac{dt}{t}\varrho(t)\frac{1}{t+\Box-i\varepsilon},$$

$$\varrho(t) = \left(1+\frac{2m^2}{t}\right)\left(1-\frac{4m^2}{t}\right)^{\frac{1}{2}}.$$

Inserted in (16.40), we get.

$$W_{wf}^{(1)}[A] = -\frac{1}{4}\int(dx)F_{\mu\nu}(x)\Pi(\Box)F^{\mu\nu}(x).$$

The total action in one-loop approximation is then written in the form

$$W = W_{Maxwell} + W^{(1)} = \int(dx)\mathcal{L}$$

with.

$$\mathcal{L}(A) = -\frac{1}{4}F_{\mu\nu}[1 - \frac{\alpha}{3\pi}\Box \int_{(2m)^2}^\infty \frac{dt}{t}\frac{\varrho(t)}{t+\Box}]F^{\mu\nu} + O(\alpha^2), \tag{16.42}$$

which is obviously gauge invariant.

Now let us choose in Eq. (16.42) $\vec{B} = 0$, $\vec{E} = const.\, in\, time \neq 0$,
so that with $-\frac{1}{4}F_{\mu\nu}F^{\mu\nu} = \frac{1}{2}\left(\vec{E}^2 - \vec{B}^2\right)$

and $-\frac{1}{4}F_{\mu\nu}[1 - \frac{\alpha}{3\pi}\Box \int_{(2m)^2}^\infty \frac{dt}{t}\frac{\varrho(t)}{t+\Box}]F^{\mu\nu} = \frac{1}{2}\vec{E}\cdot[/]\vec{E} = \frac{1}{2}\vec{\nabla}A^0\cdot[/]\vec{\nabla}A^0$,

the Eq. (16.42) yields for the electrostatic fields.

$$\mathcal{L}_{wf}^{(1)}(A) = \frac{1}{2}\vec{\nabla}A^0\cdot[1 + \frac{\alpha}{3\pi}\Delta \int_{(2m)^2}^\infty \frac{dt}{t}\frac{\varrho(t)}{t-\Delta}]\vec{\nabla}A^0. \tag{16.43}$$

Now we put $A^0 \equiv \Phi$ and define

$$V_{static} = -ext_\Phi \int d^3\vec{x}\left\{\frac{1}{2}\vec{\nabla}\Phi\cdot\left[1 + \frac{\alpha}{3\pi}\Delta \int_{(2m)^2}^\infty \frac{dt}{t}\frac{\varrho(t)}{t-\Delta}\right]\vec{\nabla}\Phi - \Phi J_0\right\}$$

$$=: ext_\Phi F[\Phi],$$

$$F[\Phi] = \int d^3\vec{x}\left\{-\frac{1}{2}\vec{\nabla}\Phi\cdot D_x\vec{\nabla}_x\Phi + \Phi J_0\right\}; \quad D_x = [1 + \frac{\alpha}{3\pi}\Delta_x\int_{(2m)^2}^\infty \frac{dt}{t}\frac{\varrho(t)}{t-\Delta}] \tag{16.44}$$

$$= \int d^3 \vec{x} \left\{ \frac{1}{2} \Phi D_x \Delta_x \Phi + \Phi J_0 \right\} \tag{16.45}$$

For the variation of $F[\Phi]$ we obtain, using the principle of stationary action,

$$\frac{\delta F[\Phi]}{\delta \Phi(z)} = D_z \Delta_z \Phi(z) + J_0(z) = 0.$$

So we find

$$D_z \Delta_z \Phi(z) = -J_0(z), \, x, z \in \mathbb{R}^3 \tag{16.46}$$

When this result is substituted in (16.44), we have

$$V_{static} = \int d^3 \vec{x} \left\{ \frac{1}{2} \Phi_{J_0} D_x \Delta_x \Phi_{J_0} + \Phi_{J_0} J_0 \right\}$$

$$= \frac{1}{2} \int d^3 x \, \Phi_{J_0} J_0 = \frac{1}{2} \int d^3 x \, J_0(x) \left( \frac{1}{\Delta_x D_x} J_0 \right)(x).$$

From (16.44) we take

$$\Delta^{-1} D^{-1} = \Delta^{-1} \left[ 1 + \frac{\alpha}{3\pi} \Delta \int\limits_{(2m)^2}^{\infty} \frac{dt}{t} \frac{\varrho(t)}{t - \Delta} \right]^{-1}$$

$$= \Delta^{-1} \left[ 1 - \frac{\alpha}{3\pi} \Delta \int\limits_{(2m)^2}^{\infty} \frac{dt}{t} \frac{\varrho(t)}{t - \Delta} \right] + O(\alpha^2)$$

$$= \frac{1}{\Delta} - \frac{\alpha}{3\pi} \int\limits_{(2m)^2}^{\infty} \frac{dt}{t} \frac{\varrho(t)}{t - \Delta},$$

and thus we find the very useful formula

$$V_{static} = -\frac{1}{2} \int d^3 x \, J_0(x) \left[ \frac{1}{\Delta} + \frac{\alpha}{3\pi} \int_{(2m)^2}^{\infty} \frac{dt}{t} \frac{\varrho(t)}{t - \Delta} \right] J_0(x). \tag{16.47}$$

Now we make use of the relations

$$(\Delta_x + k^2) \left\langle \vec{x} \left| \frac{1}{\Delta + k^2} \right| \vec{y} \right\rangle = \delta(\vec{x} - \vec{y}), \quad \left\langle \vec{x} \left| \frac{1}{\Delta + k^2} \right| \vec{y} \right\rangle = -\frac{1}{4\pi} \frac{e^{ik|\vec{x} - \vec{y}|}}{|\vec{x} - \vec{y}|}$$

$$(\Delta_x - t) \left\langle \vec{x} \left| \frac{1}{\Delta - t} \right| \vec{y} \right\rangle = \delta(\vec{x} - \vec{y}), \quad \left\langle \vec{x} \left| \frac{1}{\Delta - t} \right| \vec{y} \right\rangle = -\frac{1}{4\pi} \frac{e^{-\sqrt{t}|\vec{x} - \vec{y}|}}{|\vec{x} - \vec{y}|}$$

and obtain

$$\left(\frac{1}{\Delta - t}J_0\right)(\vec{x}) = -\frac{1}{4\pi}\int d^3y\,\frac{e^{-\sqrt{t}|\vec{x}-\vec{y}|}}{|\vec{x}-\vec{y}|}J_0(y)$$

$$= -\frac{1}{4\pi}\int d^3y\,\frac{e^{-\sqrt{t}|\vec{x}-\vec{y}|}}{|\vec{x}-\vec{y}|}\sum_i Q_i\delta^3(\vec{y}-\vec{x}_i)$$

$$= -\frac{1}{4\pi}\sum_i Q_i\frac{e^{-\sqrt{t}|\vec{x}-\vec{x}_i|}}{|\vec{x}-\vec{x}_i|}$$

$$\left(\frac{1}{\Delta}J_0\right)(\vec{x}) = -\frac{1}{4\pi}\sum_i\frac{Q_i}{|\vec{x}-\vec{x}_i|}.$$

Altogether, we end up with the formula

$$V_{static} = \frac{1}{2}\frac{1}{4\pi}\int d^3x\sum_j Q_j\delta^3(\vec{x}-\vec{x}_j)\sum_i Q_i\left[\frac{1}{|\vec{x}-\vec{x}_i|} + \frac{\alpha}{3\pi}\int\limits_{(2m)^2}^{\infty}\frac{dt}{t}\varrho(t)\frac{e^{-\sqrt{t}|\vec{x}-\vec{x}_i|}}{|\vec{x}-\vec{x}_i|}\right]$$

$$= \frac{1}{2}\frac{1}{4\pi}\sum_{i\neq j}Q_iQ_j\left[\frac{1}{|\vec{x}_i-\vec{x}_j|} + \frac{\alpha}{3\pi}\int\limits_{(2m)^2}^{\infty}\frac{dt}{t}\varrho(t)\frac{e^{-\sqrt{t}|\vec{x}_i-\vec{x}_j|}}{|\vec{x}_i-\vec{x}_j|}\right] + selfenergies \quad (16.48)$$

$$|\vec{x}_i - \vec{x}_j| = r_{ij} : V_{static} = \frac{1}{2}\sum_{i\neq j}\frac{Q_iQ_j}{4\pi}\left[\frac{1}{r_{ij}} + \frac{\alpha}{3\pi}\int\limits_{(2m)^2}^{\infty}\frac{dt}{t}\varrho(t)\frac{e^{-\sqrt{t}r_{ij}}}{r_{ij}}\right] + O(\alpha^2)$$

$$Q \equiv Q_1 = -Q_2$$

$$V_{static}(r) = -\frac{Q^2}{4\pi}\left[\frac{1}{r} + \frac{\alpha}{3\pi}\int\limits_{(2m)^2}^{\infty}\frac{dt}{t}\varrho(t)\frac{e^{-\sqrt{t}r}}{r}\right] + O(\alpha^2) \quad (16.49)$$

The second term in the square brackets of (16.49) is the well-known correction to the classical Coulomb potential. $V_{static}(r)$ was derived in the weak-field limit and thus should be valid at large distances. Because the equation of motion is linear, $V_{static}(r)$ takes the form of a superposition of Yukawa potentials.

We want to conclude this final chapter with a few remarks on the various interpretations of the term "vacuum."

In Chap. 15 we developed the intuitive notion of an effective Lagrangian as describing the dynamics of "light" fields in interaction with "heavy" fields which are hidden from direct observation. As an example, we considered fermions in an external electromagnetic field. The heavy degrees of freedom, represented by the electron (positron) field, were *integrated out*, because they are not observable. This is similar to mechanics, where one integrates out the "fast" degree of freedom in an oscillating system and discusses only the time development of the "slow" degree of freedom.

On the other hand, there is the concept of the "old" vacuum. Here, the Lamb shift in atomic levels or the anomalous magnetic moment of the electron are considered as observable consequences of the vacuum fluctuations of the photon field. This means that one gives up the concept of the vacuum being a particle- and free-field space.

This is not only true for the radiation field, but also for the electron–positron, or, more generally, for every matter field. This indicates that the vacuum is not only supposed to contain local fluctuations of the electromagnetic field strengths (which are interpreted as virtual photons), but also charge fluctuations due to the creation and subsequent annihilation of electron–positron pairs. Owing to the energy-time uncertainty principle, $\Delta E \cdot \Delta t \geq \hbar$, the maximum lifetime of such a pair is $\frac{\hbar}{2mc^2}$, where $m$ is the electron mass.

Incidentally, on the basis of Planck's and Einstein's works, Nernst—in the Berlin Academy of Sciences—proposed $\frac{1}{2}h\nu$ for the zero-point fluctuations of the electromagnetic field. Furthermore, he suggested that the entire universe is filled with zero-point energy... . We are talking about the year 1916!! Have we made any progress since then?

If we apply a sufficiently strong external electric field to the "old" vacuum, it is possible for this field to separate the electron from the positron, so that no recombination takes place. Consequently, if each particle gets an energy of at least $mc^2$ during its lifetime, $\frac{\hbar}{2mc^2} \approx 10^{-21}sec$, the *virtual* electron–positron is converted into a *real* electron (positron). It is important to notice that this creation is not a "creatio ex nihilo," because the energy corresponding to the rest mass of the created particle is extracted from the *external* applied field.

This idea of a "boiling vacuum" is certainly a useful concept that works, but is it nothing more than speculative?

The most important aspect of this vacuum model seems to be that it is endowed with an a priori structure through the presence of virtual, unobserved radiation and matter which only have to be set free from the vacuum due to an external action.

Is it not more realistic to consider the vacuum as an extended manifold, a space without metric that gets its structure through the endowment of a metric (matter) that functions as an exterior action? This is Riemann's idea presented in 1854 on the occasion of his habilitation lecture (in presence of Gauss) and finds its confirmation through the Einstein-Hilbert theory of gravity 60 years later, where space and matter are intertwined. Why not use this idea for an absolutely empty vacuum that becomes real when energy, momentum, spin and all the other attributes respecting the necessary symmetries are transferred into a certain space-time volume, thereby giving the particle its real (on-shell) existence?

That this idea works was verified in the 60s of the last century by Julian Schwinger, who formulated all of the results of quantum electrodynamics in the language of "source theory," of which we have given an example in the present chapter, where an electron-positron pair was released and absorbed by extended sources. The production of a particle pair with their masses on shell is given in Figure 16.1 by taking the imaginary part.

Last of all: Maybe the many ideas and calculations presented in the second half of this book are too abstract. Then we strongly recommend reading the "Report on Quantum Electrodynamics" that was mentioned at the beginning of chapter 12. Schwinger's contribution to this book should encourage the reader to understand how he converted from a believer in operator-to c-number quantum field theory, using as an example quantum electrodynamics. To make life easier for the interested reader, he or she can find a very enlightening excerpt from Schwinger's new assessment of the past and future development of quantum electrodynamics, which is enclosed in our following final contribution to the quantum action principle:

Renormalization, then, is the process of transferring attention from the underlying dynamical variables with which the theory begins to the physical level at which the observed particles are in evidence. It is not a concept to which there are intrinsic conceptual objections, and indeed must occur in any theory that contains structural assumptions about the constitution of the physical particles, in which abstract, underlying, dynamical variables appear.

Only now that I have described the fundamental significance of renormalization will I refer to the fact that is usually emphasized about this procedure. When one solves the coupled system of differential equations for the time-ordered field correlation functions by successive substitution, as a power series in $e_0^2$, the results contain divergent integrals. Very small distances or, equivalently, very high momenta make a disproportionately large contribution to the results, with the consequence that solutions to the coupled equations do not really exist—at least, not in the sense of a perturbative expansion. Whether this is merely a deficiency in the elementary mathematical approach, which could disappear in a more sophisticated scheme, or, an intrinsic failure of the physical model, I do not know. Nor am I very sure that it is particularly important to select one of the choices, since it must remain true that phenomena at very small distances play a substantial role in the whole story. What is important is that the renormalized equations, on the other hand, do have solutions—finite solutions, with numerical consequences that are in overwhelming agreement with experiment.

Here is something worth understanding. Somehow, the process of renormalization has removed the unbalanced reference to very high momenta that the unrenormalized equations display. The renormalized equations only make use of physics that is reasonably well established. In contrast, the unrenormalized equations critically involve phenomena in regions where we cannot pretend to know the physics. By what right do we, as physicists, claim that the laws of physics, even within the narrow domain of electrodynamics, will forever remain the same as we extrapolate to arbitrarily high energies? All experience suggests that new and unexpected phenomena will come into play. The resulting logical situation is very interesting. From the same system of equations emerge two very different attitudes. The unrenormalized description constitutes a model of the dynamical structure of the physical particles, which is sensitive to details at distances where we have no particular reason to believe in the correctness of the physics—an implicit speculation about inner structure—while the renormalized description removes these unwarranted speculations and concentrates on the reasonably known physics that is germane to the behaviour of the particles.

This way of putting the matter can hardly fail to raise the question whether we have to proceed in this tortuous manner of introducing physically extraneous hypotheses, only to delete these at the end in order to get physically meaningful results. Clearly there would be a great improvement, both conceptually and computationally, if we could identify and remove the speculative hypotheses that are implicit in the unrenormalized equations, thereby working much more at the phenomenological level. For electrodynamics, this may be just a tour de force, since it is not claimed that new numerical results will appear in this way. But, for strong interaction physics the issue is crucial. If we are caught in a formalism that has built into it implicit hypotheses, inner structural assumptions that have small chance of being correct, we shall have grave difficulties in fighting through to the correct theory. I believe that we need a more flexible kind of theory, one that can incorporate experimental results, and extrapolate them in a reasonable manner without falling into the trap of the whole-sale extrapolation that infringes on unexplored areas where surprises are sure to await. I am well aware that this kind of flexible approach is anathema to the mathematically oriented [State your axioms! What are the calculational rules?], but I continue to hope that it has great appeal to the true physicist (Where are you?).

Having said all this, let us return to electrodynamics and try to identify the implicit hypothesis, the one that has introduced speculative structural assumptions. I say that it is the introduction of operator fields. Do not misunderstand me. The field concept is unavoidable, barring some totally new approach to the space–time continuum, which is not being advocated. But operator fields? That is another matter. An operator such as $\psi(x)$ is defined by the totality, or at least a sufficiently large class, of its matrix elements. And the overwhelming proportion of these refer to energies and momenta that are far outside experimental experience. Unavoidably, then, an operator field theory makes reference to phenomena in experimentally unexplored regions. It is simple enough to identify the innocent use of field operators as the culprit. But, what to replace it with?

Usually, during the development of a particular subject, several competing but equivalent formalisms are available. The example of matrix and wave mechanics naturally comes to mind. Yet one of these may be specially suited for the transition to the next levels of description. Who, in the mid-nineteenth century could have suspected the significance of the otherwise not very useful Hamiltonian formalism in the much later developments of statistical mechanics and quantum mechanics? It is important that these variant approaches exist since, in seeking a new theory, one of them may sufficiently narrow the mental gap that needs to be traversed to make this journey feasible. The human mind is not adapted to a quantum jump in ideas. A small step for mankind is all that one can reasonably expect.

I mention this in order to recall that other formalisms for quantum electrody-namics were in existence. During the 25 year period of quantum electrodynamical development, there was great formal progress in the manner of presenting the laws of quantum mechanics, all of which had its inspiration in a paper of Dirac. This paper (which is No. 26 in the collection, Selected Papers on Quantum Electrody-namics, Dover, 1958) discussed for the first time the significance of the Lagrangian in quantum mechanics.

I have always been puzzled that it took so long to do this, but a faint glimmering of the reason appeared when I reread this paper recently and noticed that even Dirac himself thought that the action principle required the use of coordinates and velocities rather than coordinates and momenta, despite the existence of the classical action expression

$$W = \int_{t_2}^{t_1} dt \left[ \sum_k p_k \frac{dq_k}{dt} - H(p, q) \right].$$

[Incidentally, this same hang-up seems to persist in recent articles claiming that the quantum action principle is inapplicable to curved spaces.] Eventually, these ideas led to Lagrangian or action formulations of quantum mechanics, appearing in two distinct but related forms, which I distinguish as differential and integral. The latter, spearheaded by Feynman, has had all the press coverage, but I continue to believe that the differential viewpoint is more general, more elegant, more useful, and more tied to the historical line of development as the quantum transcription of Hamilton's action principle.

# References

1. Adler, S.L.: "Non-Abelian statics", talk presented at the Maxwell Symposium, Amherst (1981), Phys. Lett B **110**, 302 (1982); ibid. **113**, 405 (1982): ibid. **117**, 91 (1982); Nucl. Phys. B **217**, 381 (1983)
2. Dittrich, W., Reuter, M.: Effective Lagrangians in Quantum Electrodynamics. Springer, Berlin (1985)
3. Dittrich, W.: Some remarks on the use of effective Lagrangians in QED and QCD. Int. J. Mod. Phys. A **30**(18–19) (2015)
4. Fetter, A.L., Walecka, J.D.: Theoretical Mechanics of Particles and Continua. McGraw-Hill Book Company, New York (1980)
5. Schwinger, J., Milton, K.A., DeRaad, L.L., Tsai, W.-Y.: Classical Electrodynamics. Westview Press, Cambridge (1998)
6. Dittrich, W., Reuter, M.: Classical and Quantum Dynamics, 6th edn. Springer, Berlin (2020)
7. Weinberg, S.: Gravitation and Cosmology. Wiley, New York (1972)
8. Lenz, W.: Zs. f. Phys. **24**, 197 (1924)
9. Pauli, W.: Zs. f. Phys. **36**, 336 (1926)
10. Schwinger, J.: A note on the quantum dynamical principle. Philos. Mag. Ser. 7 **44**, 1171 (1953)
11. Dittrich, W.: The cofounder of quantum field theory: Pascal Jordan. Eur. Phys. J. H **40**, 241–260 (2015)
12. Born, M., Heisenberg, W., Jordan, P.: Zur Quantenmechanik II. Z. Phys. **30**, 558 (1925)
13. Mehra, J. (ed.): The Physicist's Conception of Nature. Reidel Publishing Company, Dordrecht (1973)
14. Born, M., Jordan, P.: Zur Quantenmechanik. Z. Phys. **34**, 858 (1925)
15. Born, M.: My Life: Recollections of a Nobel Laureate. Scribner, New York (1978)
16. Einstein, A., Born, M.: Born Einstein Letters, 1916-1955: Friendship, Politics and Physics in Uncertain Times. MacMillan Ltd., London (1971)
17. Adler, S.L.: Non-Abelian statics, talk presented at the Maxwell symposium, Amherst (1981), Phys. Lett. B **110**, 302 (1982); ibid. **113**, 405 (1982); ibid. **117**, 91 (1982); Nucl. Phys. B **217**, 381 (1983)
18. Dittrich, W.: Some remarks on the use of effective Lagrangians in QED and QCD. Int. J. Mod. Phys. A **30**(18–19) (2015)
19. Dittrich, W., Reuter, M.: Effective Lagrangians in Quantum Electrodynamics. Springer, Berlin (1985)

# Further Reading

1. Dittrich, W.: On Schwinger's various approaches to quantum field theory. Fortschritte der Physik **22**, 263–293 (1974)
2. Dittrich, W., Holger Gies, H.: Probing the Quantum Vacuum, Perturbative Action Approach to QED and Its Applications. Springer Tracts in Modern Physics, vol. 166. Springer, Berlin (2000)
3. Dittrich, W.: The Heisenberg-Euler Lagrangian as an example of an effective field theory. Int. J. Mod. Phys. A **29**(26) (2014)
4. Feynman, R.P., Hibbs, A.R.: Quantum Mechanics and Path Integrals. McGraw-Hill, New York (1965)
5. Fried, H.: Functional Methods and Models in Quantum Field Theory. The MIT Press, Cambridge (1972)
6. Gozzi, E., Pagani, C., Reuter, M.: The response field and the saddle points of quantum mechanical path integrals (2020). arXiv:2004.08874v1
7. Horn, B.K.P.: The curve of least energy. Technical report A.I. Memo. No. 612, MIT AI Lab (1981)
8. Love, A.E.H.: A Treatise on the Mathematical Theory of Elasticity, 4th edn. p. 401. Dover Publications, New York (1944)
9. Mehlum, E.: Non linear splines. Computer Aided Geometric Design, pp. 173–205 (1974)
10. Milton, K.: Schwinger's Quantum Action Principle. SpringerBriefs in Physics. Springer Nature, New York (2015)
11. Schwinger, J.: On gauge invariance and vacuum polarization. Phys. Rev. **82**, 664 (1951)
12. Schwinger, J.: Particles, Sources and Fields, vol. 1 (1970); vol. 2 (1973); vol. 3 (1989). Addison-Wesley Publishing Company, New York
13. Toms, D.J.: The Schwinger Action Principle and Effective Action. Cambridge Monographs on Mathematical Physics. Cambridge University Press, New York 2007 (2012)

© The Author(s), under exclusive license to Springer Nature Switzerland AG 2021          135
W. Dittrich, *The Development of the Action Principle*,
SpringerBriefs in Physics,
https://doi.org/10.1007/978-3-030-69105-9

Printed in the United States
by Baker & Taylor Publisher Services